Canada's Boreal Forest

SMITHSONIAN NATURAL HISTORY SERIES

John Kricher, Series Editor

Books in this series explore the diverse plants, animals, people, geology, and ecosystems of the world's most interesting environments, presented in an accessible style by world-renowned experts.

Canada's Boreal Forest

J. David Henry

SMITHSONIAN
INSTITUTION
PRESS

WASHINGTON
& LONDON

© 2002 by the Smithsonian Institution
All rights reserved

Copy editor: Diane Hammond
Production editor: Robert A. Poarch
Designer: Brian Barth

Library of Congress Cataloging-in-Publication Data
Henry, J. David, date.
　　Canada's boreal forest / J. David Henry.
　　　　p. cm. — (Smithsonian natural history series)
　　ISBN 1-58834-057-0 (alk. paper)
　　1. Taiga ecology—Canada. 2. Taigas—Canada. I. Title. II. Series.
QH106.H46 2002
577.3′7′0971　　　　　　　　　　　　　　　　　　　2001057668

British Library Cataloguing-in-Publication Data is available
Manufactured in the United States of America
09　08　07　06　05　04　03　02　　5　4　3　2　1

Color signature printed in China.

♾☼ The recycled paper used in this publication meets the minimum requirements of the American National Standard for Information Sciences—Permanence of Paper for Printed Library Materials ANSI Z39.48-1984.

For permission to reproduce the photographs appearing in this book, please correspond directly with the owners of the works, as listed in the individual captions. (The author owns the photographs that do not list a source.) The Smithsonian Institution Press does not retain reproduction rights for these photographs or illustrations individually or maintain a file of addresses for photo sources.

To my family and friends

*Together we always have found
an inch or two of freeboard
while crossing the big openings of life.*

May we continue to do so.

Contents

	Editor's Note	ix
	Preface	xi
	Acknowledgments	xv
1	The Sweep of the Taiga	1
2	A Northern Primer	9
3	Knobs, Kettles, and Precambrian Corks	25
4	Seasons to Burn	39
5	The Benefits of Being Burned	47
6	A Forest in Search of a Fire	59
7	The Taiga in Winter	69
8	A Hare's Breath from Death	85
9	Muskegs—Halfway between Land and Water	101
10	Northern Lakes, Troubled Waters	115
11	The Conservation of the Uncommon Loon	129
12	The Land That God Gave Cain	141
13	The Nordic Challenge	151
14	Places to Visit in Canada's Boreal Forest	165
	Index	173

Editor's Note

Circling the high latitudes of Earth is a belt of conifers, mostly spruces and firs, that forms the basis of one of the world's most expansive and fascinating ecosystems, the boreal forest. Exposed to short days and heavy snowfall during the long northern winter and subjected to a decidedly compressed summer growing season, the plant species of the boreal forest have adapted. Tough evergreen needlelike leaves and a conical shape that easily sheds winter snows are but two of many adaptations that permit the spruce and the fir to prevail where most other tree species would perish. Coexisting with them are many species of shrubs and wildflowers, some confined to acidic bogs, themselves remnants of recent glaciation, another unique characteristic of the boreal landscape. Sometimes called the *spruce-moose biome,* the boreal forest is an ecosystem where the annual rhythms are very closely attuned to the changing seasons. In summer there is a flush of insects, fed upon by millions of long-distance migrant birds that winter in the American tropics but that annually fly thousands of miles north with the warm winds of spring to breed in the dense forest.

The boreal forest, sometimes called the taiga, is in some ways an environment in climatic transition, bordering temperate hardwood forests and prairies to the south and vast, sweeping, treeless Arctic tundra to the north. Among the many muskeg bogs and black spruce swamps that dot the broad landscape are uniform forests of white spruce and balsam fir, occasionally interrupted by stands of quaking aspen or jack pine, species that are among the first to thrive after the occurrence of natural fire. Moving northward, the trees diminish in size and abundance, as the spruces and firs yield to dwarf willows, and soon tundra grasses and lichens come to prevail. Treeline is a ragged boundary that attests to the ongoing struggle between organisms and the physical environment.

The boreal forest is best seen in the Western Hemisphere in Canada, where it is the dominant environment in almost all of the Canadian provinces. It is a land where summer nights are punctuated by the haunting yodel of common loons, where the night skies are illuminated by the aurora borealis, and where packs of timber wolves stealthily hunt moose. It is a land of geology as well as natural history, home to some of North America's oldest rock formations. It is a land where natural fires play a fundamental role in maintaining biodiversity.

J. David Henry is ideally qualified to be your guide to understanding Canada's boreal forest. David, an accomplished and widely published boreal ecologist, holds the position of conservation ecologist for Parks Canada in the Yukon. He has traveled both within the North American boreal forest as well as its counterpart in northern Europe so has a unique and comprehensive understanding of the region's natural history. David is the author of *Red Fox: The Catlike Canine* (Smithsonian Institution Press, 1996), a book that clearly demonstrates his talents as a writer as well as his feeling for the remarkable animal that is his subject. Now he expands his focus to give attention to the sweeping natural history of the boreal forest. Put on your intellectual snowshoes and walk with David in the land of lynx and wolverine.

John Kricher

Preface

The taiga, also known as the boreal forest, is a land deeply loved but little understood. Across North America each summer, families seek out the boreal forest for boating, swimming, and relaxing at its unspoiled lakes. Others canoe on the Canadian Shield or come to the taiga each spring for its unparalleled sportfishing. I wonder how many of these vacationers are aware that we are participating in cultural traditions spanning generations and shared by several circumpolar cultures.

Consider the cabin. The family cabin on a boreal lake or in the woods is a deeply ingrained part of northern cultures, indigenous and nonindigenous alike. It is the place where we share special moments and cherished family traditions. It is where we go to rest, to reflect, and to get in touch with nature. Often the cabin has been in a family for generations. We may travel hundreds or even thousands of miles each summer to return to it. In Scandinavia it is where the ashes of parents and grandparents are often spread. For many extended families, allowing ownership of the cabin to leave the family would be like selling the clan's iconic home. In North America the cabin may also be called the *cottage* or the *camp;* in Quebec it is the *country house* or *chalet;* in Russia it is the *dacha;* in Sweden, the *forest cottage,* and in Finland, *mökki*. In all these northern cultures the family cabin is an important link between people and the land. Whether it is time at the cabin, the summer canoe expedition, a backpacking trek, or a cherished fishing trip, we value our time in the boreal forest. It is an all too brief time of reflection and vigor, which enriches our lives.

In spite of the affection that we northerners have for our boreal forest, most of us abandon it once summer has passed and snow covers the ground. We forget it for another year and allow it to become a land of clear-cuts, mining projects, and water-diversion schemes. Why do we surrender this land so easily? Perhaps it is our lack of understanding of the great northern forest that limits our commitment to it.

It is not surprising that our comprehension of the boreal forest is somewhat fragmentary. It is one of the largest biomes (i.e., life zones) on Earth. In Canada alone, it covers more than 1.5 million square miles (approximately 4 million km^2), fully

The lakeside cabin is an important tradition in many northern cultures.

one-third of the country. However, it is also a harsh and remote land. Minus 40° temperatures characterize the winter, and hungry hordes of mosquitoes and black flies are constant companions during the summer. It is a harsh land in which to live and work. As a result of the region's remoteness and severity, the scientific community has carried out only limited research on its ecology. We clearly understand some attributes and certain ecological processes occurring in the taiga, but there are many holes in our knowledge of boreal ecology. This book attempts to summarize what we understand about its ecology, but the attentive reader will sense information gaps and the hundreds of questions that remain unanswered about this northern land. Our understanding of the taiga is fragmentary at best.

My purpose in writing this book is twofold. First, I hope to increase our understanding and enjoyment of the northern woods. Second, I hope to challenge the commonly held perception of the land as an immense economic barrens where highly exploitive resource schemes are not only tolerated but also often heavily subsidized. Furthermore, such views of the taiga are centuries old; the early explorers referred to the northern forest as "the land of little sticks." Let me illustrate these points with two examples.

In the late seventeenth century, a group of London businessmen formed the Hudson's Bay Company to exploit the rich fur resources of northern Canada. On May 2, 1670, King Charles II of England granted this "company of adventurers" a royal charter giving them complete control of all the territory drained by the rivers flowing into Hudson Bay. This huge domain, called Rupert's Land after the king's cousin, encompassed nearly 3 million square miles (7.7 million km^2), and the king of England simply gave away complete control of this vast land, free of charge. The king received no taxes and no royalties as part of this charter.

My second example comes from more recent times. Since 1989 an area of Alberta nearly the size of New England has been committed almost exclusively to the production of wood pulp. In that year, as a result of a few short months of negotiations, the government of Alberta signed several forest management agreements committing more than a quarter of the province—over 64,000 square miles (166,000 km^2) of Alberta's boreal forest—to the production of wood pulp. There is no provision in the forest management agreements for the conservation of wilderness, the establishment of parks, the preservation of old-growth forests, or the settlement of land claims with the Lubicon Cree First Nation or other native groups.

The Daishowa forest management agreement is typical. The government offered Daishowa Corporation (a group of Japanese businessmen) a taxpayer-financed subsidy of close to $75 million to build two mills for bleached kraft pulp and to harvest enough wood fiber to run these pulp mills at near capacity. The agreement guaranteed Daishowa exclusive rights to more than 15,300 square miles (41,000 km^2) of boreal forest in northern Alberta. The corporation did not have to pay the government of Alberta royalties for a number of years. Economic development and the new jobs created in northern Alberta are the only benefits that Albertans have received.

These two examples span three centuries, but the perception of the boreal forest as an economic wasteland—a giveaway land—is the same. In fact, the perception of the taiga may have deteriorated; Charles II at least broke even.

Is this the only way to view the Canadian taiga? Consider, by contrast, the ecologically similar boreal forest in Sweden: known as the Swedish National Forest, it is viewed as every Swede's legacy and one of the foundations of the Swedish culture. And although the taiga in Siberia used to be considered a giveaway kingdom, during the twentieth century it became one of the foundations of Russia's economy.

The focus of this book is the boreal forest in Canada. However, the boreal forest is a circumpolar ecosystem, and it is ecologically similar around the globe. Thus to understand the Canadian taiga, it is useful to refer to research carried out in the boreal forest of Alaska as well as in the northern woods of Minnesota and Maine and to compare Canada's taiga to the taiga in northern Scandinavia, Finland, and Russia—with the hope that these comparisons will help place the snow forest of Canada in its proper context.

J. David Henry

Acknowledgments

This book relies on the research and writings of many scientists. Some of my sources are listed in the references given at the end of each chapter. Discussions with scientists, naturalists, aboriginal people, and writers were also helpful. Some of these people are acknowledged below. The format of this series is not compatible with a detailed acknowledgment of all of my sources; however, a study guide, described below, gives a more detailed listing. To all who assisted in the development of this book and the study guide, I express my thanks and gratitude.

Dr. William O. Pruitt Jr. has been supportive of the project since its inception and has helped in innumerable ways. Dr. William A. Fuller read an early version of the manuscript, and his comments on scientific content and writing style assisted me greatly. Other people who read the manuscript as it developed include Dr. J. Stan Rowe and Dr. Wayne Lynch. Others critiqued one or several chapters, and for their help I wish to thank Thomas Berger, Dr. Anders Bjarväll, Dr. Rudy Boonstra, David Dodge, Dr. David Gauthier, Dr. Scott Gilbert, Dr. David Hamer, Stuart Heard, Elizabeth Henry, Dr. Mary Henry, Alice Kenney, Dr. Charles J. Krebs, Torbjörn Lahti, Dr. John Lewry, Ingemar Lext, Dr. Erik Lindström, Candace Savage, Dr. David W. Schindler, Lindsay Staples, and Dr. John Theberge.

Other people shared their insights about the boreal forest or assisted in other ways. For their help, I wish to thank Marilyn Brass, Dr. John Bryant, Phil Caswell, and Dr. Dennis Dubé; Ken East and everyone from the Parks Canada Yukon Field Unit; Agneta Enebro, Reidar Erbe, John Hastings, and Elizabeth Hofer; Dr. Elina Helander and Lars-Nila Lasko of the Nordic Sami Institute; Dr. Stephen Herrero, Alf Isak Keskitalo, Ken Kingdon, Anne Landry, Ingemar Lext, Erik Lindström, Michelle Mico, Bradley Muir, Erik Myrhaug, Py Nasman, Sergei Nazaroff, Kerstin Persson, Dr. Erkki Pulliainen, Laud Ruoftinen, Dr. David Scott, Dr. George Scotter, Helena Seppänen, Oystein Steinlien, Dr. Carl Tamm, Dr. Erik Wirén, Ulrich Wotschikowsky, and Dr. Olle Zackrisson. Discussions with Dr. John F. Lewry of the University of Regina about Precambrian geology were most helpful. Dr. Ahab Spence and Solomon Ratt of the Saskatchewan Indian Federated College provided standard Roman orthography for the Cree words I use. Harold and Meta Johnson allowed me to photograph traditional structures at Kwaday Dan Kenjii near Champagne, Yukon Territory. Prince Albert National Park and Riding Mountain National Park contributed photographs; in return, I hope that this book contributes to their impressive interpretive programs. Ross Barclay also contributed photographs. The other photographs are mine. Sharon Clark and Mike St. Pierre assisted with the bibliography in the study guide. To all of these people and organizations I express my gratitude.

Dr. Peter Cannell of the Smithsonian Institution Press and Dr. John Kricher, editor of the Smithsonian Natural History Series, supported the project through all of its stages. The science acquisitions editor, Dr. Vincent Burke, streamlined the format of the series and helped to make it a reality. Nicole Sloane assisted with photographs and maps, and Diane Hammond provided insightful editing. My thanks to all at the Smithsonian Institution Press for their assistance.

Those wishing to use this book as a textbook or as a resource for northern research may wish to procure the study guide, which includes supplementary information about each chapter, identifies references and sources used in each chapter, and presents an extensive bibliography on the boreal forest. Please contact me through the Smithsonian Institution Press or directly at the following e-mail address: canadasborealforest@yahoo.ca. Comments and reactions concerning the book are welcomed.

My wife, Suzanne, has provided skillful editing for all my published writings, and this book is certainly no exception. Our discussions and her insightful editing of each chapter were important contributions. Living with a writer in the house is hard on any family, as mine can readily attest. I am grateful to Suzanne and our daughter, Elizabeth, for their understanding and support. I value all that we have shared and discovered together while living in the Canadian taiga.

1
The Sweep of the Taiga

Early in the twentieth century Anton Chekhov, the great Russian playwright, made an expedition to the taiga of Siberia, traveling most of the way between Moscow and the island of Sakhalin on horseback. He was moved, and sometimes terrified, by the vastness of the taiga. Reflecting back on his trip, he later wrote: "The strength and charm of Siberia does not lie in its giant trees and its silence, like that of a tomb, but in that only migratory birds know where it finishes. On the first day one does not take any notice of the taiga; on the second and third day one begins to wonder, but on the fourth and fifth day one experiences a mood as if one would never get out of this green monster."

The taiga is immense. Rolling westward from the island of Newfoundland, it sweeps across the North American continent in a wide, uninterrupted, evergreen swath. Skipping over the Bering Sea, the forest straddles the eleven time zones of Russia and covers most of Finland and the northern two-thirds or more of Scandinavia. From a satellite orbiting high above Earth, the taiga appears as a dark mantle draped across Earth's shoulders, a robe glistening with aquamarine lakes. This forest-green cloak declares to the rest of the solar system that this planet is the home of living things.

The terms *taiga, boreal forest, snow forest,* and *northern coniferous forest* all refer to Earth's northernmost forest region. If forms a broad belt, predominantly of evergreen trees, that encircles the globe. In some parts of North America it spans more than 1,250 miles (2,000 km) from north to south, from the Mackenzie River delta to the mountains of southern Alberta and northern Montana. In central Asia it reaches even farther, from near the Arctic coast southward into the northern regions of Mongolia, a distance of more than 1,850 miles (3,000 km).

Biomes, the eleven major terrestrial life zones on the planet, are distinguished by each having its own distinct fauna and flora (with some overlap) as a result of the distinct climate of the biome. The grassland prairie, Arctic tundra, tropical rainforest, and desert are examples of biomes. The boreal forest, one of the largest biomes, occurs south of the Arctic tundra and occupies the full width of North America and Eurasia, or more than 5 million square miles (nearly 13 million km^2). Nothing similar to the taiga occurs in the Southern Hemisphere.

Let's look at the origins for the three names that are commonly used for this biome. *Taiga* (pronounced TIE-gah) is a Russian word originally used to refer to the dense, unbroken evergreen forests of the North or of Siberia. It is a common term, used by Russians in everyday language. Interestingly, *taiga* is not a Slavic word. One theory derives it from the Turkic root *dag,* meaning mountain. As the Slavic groups who became the Great Russians expanded northward and eastward during medieval times, they picked up Finnic, Ugric, and Turkic terms to describe the northern world that surrounded them. *Taiga* is one such term. When *taiga* came into use as a scientific term in the West, some ecologists restricted its meaning to the open parklike forest just south of treeline. This is clearly not the Russian meaning of the word. Today for most Western scientists *taiga* refers to the entire boreal forest, the entire biome, from the northern treeline south to where it gives way to grassland prairie, hardwood forest, and other biomes.

Boreal (pronounced BORE-ee-al) comes from the Latin word *boreas,* which refers to the Greek god of the north wind. *Boreal* means something related to or located in northern regions. Boreal forest, then, is the northern forest; the term also refers to the entire biome. *Coniferous* refers to conifer trees—the dominant, but certainly not the only, tree species in the boreal forest. Conifers are plants with needles as their leaves and cones as their seed bodies. As a taxonomic term, *conifer* refers to plants of the division Coniferophyta. In this book, *taiga, boreal forest,* and *northern coniferous forest* are used interchangeably to refer to the immense, circumpolar, northern evergreen forest.

The boreal forest in North America stretches from Alaska and the Rocky Mountains eastward to the Atlantic Ocean. Across North America, in a few northern locales, the taiga is bordered by marine areas—for example, around James Bay, the southern part of Hudson Bay, and parts of Labrador and Newfoundland. However, the taiga is usually bounded on the north by the Arctic tundra. On the south, it is bounded in the East by deciduous hardwood forests and in the West by the aspen parkland, a transition zone that gives way to the prairie grasslands. Farther west the taiga inhabits the northern portion of the great arc of western mountains, the North American Cordillera. Forests of conifer trees commonly grow on the slopes of these mountains, and it is difficult to define precisely where the boreal forest ends and the montane coniferous forest begins. The boreal forest in the western mountains is characterized by black spruce *(Picea mariana),* white spruce *(Picea glauca),* white, or paper, birch *(Betula papyrifera),* and balsam poplar *(Populus balsamifera).* The montane coniferous forest is characterized by Engelmann spruce *(Picea engelmannii),* Douglas fir *(Pseudotsuga menziesii),* and Alpine fir *(Abies bifolia).* Moisture increases as we move westward, because of the influence of the Pacific, so a temperate coastal rainforest predominates along the west coast. This forest is characterized by western red cedar *(Thuja plicata),* western hemlock *(Tsuga heterophylla),* Sitka spruce *(Picea sitchensis),* and yellow cypress *(Chamaecyparis nootkatensis).*

The montane coniferous forest extends into Central America, where it takes the form of high-elevation, tropical pine forests. Biogeographers who map on a global scale consider the boreal forest and these montane coniferous forests as one biome—the coniferous forest. Other biogeographers, mapping at a finer, continental scale, identify as separate biomes the moist coniferous forest of the Pacific coast, the montane coniferous forest of the central and southern Rocky Mountains, and the boreal forest of Alaska and Canada.

Further complications with classification are encountered in mountainous regions. The ecologist Linda Kershaw points out that the varied environmental conditions in the mountains cause great variation in vegetation, particularly in the western mountains, where environments change dramatically with changes in elevation and latitude. The low angle of the Sun in northern regions keeps north-facing mountain slopes much shadier and thus much moister and cooler than south-facing slopes. As a result, different vegetation frequently grows on opposite sides of a mountain.

Because of all this variation, it is often more useful to think in terms of vegetation zones, rather than biomes, in mountainous regions. As a general rule, a rise of 325 feet (100 m) in elevation causes the same changes in vegetation that one would observe traveling north approximately 185 miles (300 km). In the Canadian Rockies four vegetation zones are generally recognized. The foothills zone consists mainly of lodgepole pine *(Pinus contorta)* and trembling aspen *(Populus tremuloides)*, with understories dominated by shrubs such as soapberry *(Sheperdia canadensis)*, prickly rose *(Rosa acicularis)*, wild red raspberry *(Rubus idaeus)*, and common bearberry, or kinnikinnick *(Arctostaphylos uva-ursi)*. South of Edmonton the foothills zone grades into the prairies to the east and into the intermontane plain to the west. North of Edmonton it grades into the boreal forest east and west.

The montane zone supports a great variety of trees, shrubs, and wildflowers and varies from dry, open forests at low elevation to dense forests as it merges with the subalpine zone. Indicators of the montane zone are the presence of Douglas fir in the southern Canadian Rockies and cooler, damper forests of lodgepole pine, trembling aspen, and balsam poplar in the northern Canadian Rockies.

The subalpine zone extends from the upper edge of the montane zone to the lower edge of the treeless Alpine zone and includes stands of stunted, twisted trees (Krummholz trees), which extend into Alpine meadows at high elevation. The zone consists typically of subalpine fir and Engelmann spruce, with a ground cover typically of blueberries *(Vaccinium* species), low bush cranberry *(Viburnum edule)*, crowberry *(Empetrum nigrum)*, Labrador tea *(Ledum groenlandicum)*, soapberry, and diverse mosses.

The Alpine zone extends from treeline to the end of the vegetation on exposed slopes or ridges or to permanent snowfields. The Alpine tundra supports a variety of sedges, herbs, lichens, and ground shrubs. Its wildflowers—among them larkspur *(Delphinium* species), milk-vetch *(Astragalus* species), and Indian paintbrush *(Castilleja* species)—during June and early July in national parks such as Jasper, Glacier, and Revelstoke are truly spectacular.

As mentioned, the boreal forest in the mountains only gradually changes into the montane coniferous forest to the south and the moist coniferous forest toward the Pacific coast. However, sometimes we are forced to draw lines on maps. The map shows where experts normally draw the boundaries of the boreal forest in North America, the boundaries that this book also recognizes.

Variations on a Theme

The boreal forest covers approximately 28 percent of the North America continent north of Mexico. A number of classification schemes have been put forward to make

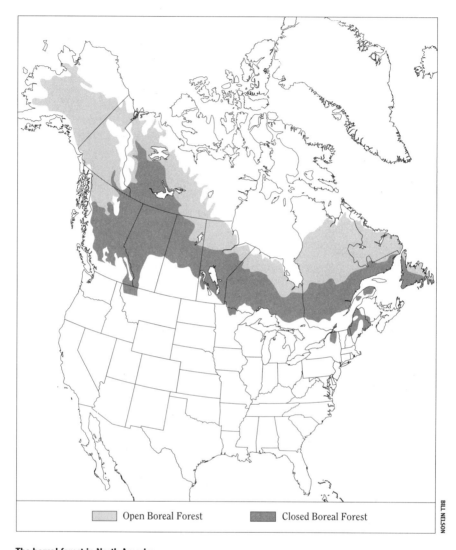

The boreal forest in North America

sense of its regional variation. In this book, I keep things simple and classify the boreal forest into one of three forms. In approximately the southern half is a fairly dense forest, its defining feature being that the trees grow close enough together and have large enough branches to form a closed canopy overhead, creating deep shade. Thus the southern half of the taiga is known as the *southern boreal forest,* or the *closed boreal forest*. The northern half of the taiga, or the *northern boreal forest,* is referred to as the *open boreal forest* because its trees grow far enough apart that a canopy never forms. Sunlight and snow cover the ground around each tree or clump of trees. The northern boreal forest is also called the *lichen woodland*. Farther north still, the ecotone (a transition zone between two biomes) between the boreal forest and the Arctic tundra is referred to as the *forest tundra* ecotone, or the *tundra coniferous forest* ecotone. It is described in more detail in the next chapter. Throughout this book I use this three-part classification of the taiga: the southern, or closed, boreal forest; the northern, or open, boreal forest; and the forest tundra ecotone.

Underlying much of the taiga is permafrost (permanently frozen ground). Technically speaking, permafrost is rock, soil, or organic matter, with or without moisture or ice, that has remained colder than 32°F (0°C) continuously for more than two years. Often it has been continuously frozen for hundreds or thousands of years. A layer of continuous permafrost underlies all of the forest tundra and part of the northern boreal forest. Under the rest of the northern boreal forest and a good portion of the southern boreal forest is a layer of discontinuous permafrost. In the North a permafrost layer normally forms fairly close to the surface of the ground (from a few inches to less than 10 feet [7 cm to 3 m] below the surface); its thickness ranges from only a thin layer to more than 300 feet (90 m). Interestingly, those lakes and rivers in the North deep enough or swift enough not to freeze to the bottom usually have no permafrost beneath them. Permafrost is a significant force in the taiga. It chills the soil above it and slows decomposition; it forms an impermeable layer, retarding drainage; and its upper layer (the active layer) goes through cycles of freezing and thawing, creating palsas, frost plateaus, and other land forms (see Chapter 9).

The taiga extends farther northward in the West than in the East, reaching into the Mackenzie River delta, a latitude of almost 70°. In the West the closed boreal forest also extends farther north, reaching the southern shores of Great Bear Lake. Both of these patterns stem from the climate, which is moderated by warm westerly winds. Northern Scandinavia has similar patterns, these being due to the Gulf Stream, which originates in the Caribbean, surges across the Atlantic, and dumps 8 million gallons (30 million liters) of relatively warm water each hour along the Norwegian coast. This major ocean current and its trade winds moderate the climate of Scandinavia and allow the taiga to grow as far north as 71°, a latitude corresponding to approximately the center of the Greenland ice cap on the other side of the Atlantic.

The People of Canada's Boreal Forest

According to the 1995 Canadian census, approximately 3,900,000 people, or 14 percent of its population, live in the boreal forest, which covers 35 percent of Canada's land area. The forest's most populated region is the boreal shield ecozone (so termed by Environment Canada in its 1996 report), a region of the Precambrian shield extending from northeastern Saskatchewan south of James Bay, eastward through Quebec, and including the island of Newfoundland. Approximately 2,832,000 people reside in this area. Major population centers of the boreal shield ecozone include the cities of St. John's, Chicoutimi, Rouyn-Noranda, Timmins, Sudbury, Sault St. Marie, and Flin Flon. The least populated region in Canada's boreal forest is the area that Environment Canada terms the *tundra cordillera ecozone,* which occupies the northern half of the Yukon Territory. Approximately 300 people reside in this area, 80 percent of whom live in the town of Old Crow, a settlement accessible only by air and water.

The density of the human population in most of Canada's boreal forest region is low, averaging less than 2.6 persons per square mile (1 person per km^2). However, there

Trees in the closed boreal forest grow close enough together to form a closed canopy during summer, blocking most of the sunlight from reaching the forest floor.

are some exceptions. In the Peace River district of northern British Columbia and in the Lesser Slave Lake region of northern Alberta, population density approaches 130 persons per square mile (50 persons per km^2).

Canada's aboriginal people historically occupied every region of the country and every part of the boreal forest. Today the subarctic First Nations occupy a vast area of the taiga, stretching from Labrador to the Yukon. Their seasonal movements and large-scale migrations during prehistory and historic times are only partially understood. For example, certain groups, such as the Chipewyan, once moved onto the tundra in certain seasons. Other groups, such as the Woods Cree, migrated into new areas once contact with Europeans had been made.

These aboriginal peoples encountered Europeans over an extended period, beginning with the Beothuk of Newfoundland, who made contact with the Vikings in approximately 1000 C.E. L'Anse aux Meadows National Historic Site, on the North Peninsula of Newfoundland, commemorates the Viking settlement of Vinland and includes several reconstructions of Norse communal sod houses. The last aboriginal groups to make contact with Europeans were probably the Tagish, the Interior Tlingit, and the Southern Tutchone, who encountered the Klondike gold rushers in southwestern Yukon around 1900 C.E. These events are commemorated in the Chilkoot Trail National Historic Park and Kluane National Park and Reserve.

The lives of the aboriginal groups changed greatly as they traded with Europeans throughout Canada's fur-trading regions and with Russians in Alaska and along the west coast of Canada. Great numbers of First Nations people died as a result of exposure to European diseases, to which they had no immunity. The Beothuk, for example, died out completely during the mid-1800s because of, among other factors, European diseases. Many aboriginal groups gradually recovered, however, and today these First Nations maintain their distinct cultures, languages, and activities in their traditional territories.

The native peoples of Canada's taiga speak languages from two broad families of aboriginal languages: Algonquian-based languages in the central and eastern regions of Canada and Athapaskan-based languages in the western and northern regions. Cree, Montagnais-Naskapi, Ojibwa, and Saulteaux are Algonquian-based languages. Hare, Han, Gwitchin, and Tutchone are Athapaskan-based languages. Tlingit is not related to either linguistic family and is spoken only by the Tlingit groups of the Northwest.

The snow forest is inhabited by aboriginal people across its whole expanse, and it is their homeland that we are entering when we come to visit. Thomas Berger, who has chaired a number of public hearings in Alaska and northern Canada, observes, "It is not always easy to remember when one is flying over the lake-speckled, unlimited greenness of the boreal forest that the region has been inhabited by First Nations for many thousands of years. The human populations that have used this great area were never large, but their skills as travelers and hunters made it possible for them to occupy virtually all of the land." In his book *Village Journey* Berger quotes aboriginal people who attended his public meetings in Alaska. Many mentioned the strong connection between the land and their culture. Walter Soboleff, of Tennakee, Alaska, said "No matter what the weather may be like, to know that we own land gave us comfort, gave us refuge. It was home. From it, we gained food. From it, we gained medicine. On it, we performed the ancient ceremonies. It gave strength to the clan, it gave strength to the family life, and courage and pride to carry on their way of life." And Polly Koutchak, of Unalakleet, Alaska, made this observation: "How firm we stand and plant our feet upon our land determines the strength of our children's heartbeats."

General References

Canadian Geographic. 1996. *The Boreal Forest: The National Atlas of Canada.* Ottawa.

Environment Canada. 1996. CD ed. *Conserving Canada's Natural Legacy: The State of Canada's Environment—1996.* Ottawa: Environment Canada.

Freeman, M. M. R., ed. 1981. *Proceedings of the First International Symposium on Renewable Resources and the Economy of the North.* Ottawa: Association of Canadian Universities for Northern Studies.

Fuller, W. A., and J. G. Holmes. 1972. *The Life of the Far North.* New York: McGraw-Hill.

Johnson, D., L. Kershaw, A. MacKinnon, and J. Pojar. 1995. *Plants of the Western Boreal Forest and Aspen Parkland.* Edmonton: Lone Pine Publishing.

Kershaw, L., A. MacKinnon, and J. Polar. 1998. *Plants of the Rocky Mountains.* Edmonton: Lone Pine Publishing.

Larsen, J. A. 1980. *The Boreal Ecosystem.* New York: Academic Press.

———. 1982. *Ecology of the Northern Lowland Bogs and Conifer Forests.* New York: Academic Press.

Marsh, J. H., ed. 1994. *The Canadian Enclyclopedia*, Edmonton: Hurtig Publishers.

Pruitt, W. O., Jr. 1978. *Boreal Ecology.* Studies in Biology 91 (Institute of Biology). London: Edward Arnold Publishers.

2
A Northern Primer

The boreal forest occupies 35 percent of the land area of Canada. It is generally dominated by jack and lodgepole pine, black and white spruce, balsam fir, and tamarack, with the dominant conifer varying regionally and with local habitat. Broad-leaved trees, such as aspen, balsam poplar, and paper birch, occur either in pure stands or mixed with conifers. These broad-leaved species become increasingly dominant toward the southern edge of the boreal forest. Species diversity is low in the North. These nine tree species are the only ones that are widespread throughout the North American boreal forest. They are the hardy ones, the only tree species that can cope with the harshness of boreal winters.

During the course of this book, I mention places you can visit to experience the boreal forest in Canada. Canadian national parks are highlighted because these are some of the most accessible and yet best protected regions of the Canadian taiga. National parks are also among the finest locations to observe the undisturbed behavior of wild animals in their natural taiga setting. (Information on the national parks, provincial parks, museums, cultural centers, and other locations mentioned in this book is given in Chapter 14.)

This book focuses on the Canadian taiga, but the taiga also covers more than half of Alaska, most of northern Maine, northern New Hampshire, some of the higher elevations in Vermont, and the Adirondack Mountains in upper New York State (see map in Chapter 1). It invades northern Minnesota and blends with the montane conifer forest that is found in western Montana, Idaho, Washington, Alberta, and British Columbia. So while the book is a guide to the Canadian taiga, the ecology that it explains is largely valid for these other regions of the North America taiga.

Seasonal Extremes

If the Earth were a perfect sphere with no clouds circulating above its surface, and if its axis (the line through the North and South Poles) did not tilt relative to the Sun, every point on the surface of the Earth would receive the same hours of sunlight each year. In fact, the tilting does not affect the total number of hours of sunlight delivered each year but only the seasonal pattern in which it is delivered. As the Earth on its tilted axis makes its annual rotation around the Sun, the northern end of the Earth is tilted toward the Sun in summer, making the Northern Hemisphere warmer, and away from the Sun in winter, making it cooler. At the equator, the same hours of sunlight are received 365 days each year. In extreme northern and southern regions the winters are dominated by darkness and the summers by almost continuous daylight. At the Arctic Circle (66°30′) there is one day during summer when the Sun never sets and one day during winter when the Sun never rises.

Given the abundance of sunlight received during the growing season in the North, why isn't the Arctic lush and green like the Tropics? One factor that makes a major difference between the North and the Tropics is the angle at which the incoming sunlight strikes the surface of the ground. In the Tropics the sunlight comes in from overhead, striking the ground at nearly a perpendicular angle for much of the year. Because of the Earth's spherical shape and tilt, closer to the poles the Sun appears low in the sky, and the sunlight strikes the Earth not perpendicularly but at a low angle. Anyone who has shingled a roof where on one side the Sun is striking the roof nearly perpendicular and on the other side at a much lower angle has experienced the difference that the angle of incidence can make.

Sunlight coming in at a low angle affects the quality of solar energy received in two ways. First, much more energy is reflected off the surface of the ground than is absorbed by it, causing the ground to be colder and vegetation communities to be less productive than in tropical regions. Second, because sunlight passes through more atmosphere in the North than in the Tropics, dust, water vapor, and other microscopic particles in the air absorb or reflect a much higher percentage of the incoming solar radiation. Thus, even though northern regions have long hours of daylight or even continuous daylight near the summer solstice, the quality of that solar energy is drastically reduced by the angle of the Sun.

As the North passes from summer into winter, the region develops a serious energy deficit. In areas north of 40° latitude, an increasingly greater amount of incoming energy either is transferred from the surface of the snow or ground to the atmosphere or is simply reflected back to outer space, compared to the amount of energy gained. During summer, all regions receive a surplus solar energy. However, as the Canadian ecologist Bill Pruitt shows, during winter northern areas fall into a negative energy balance. The farther north you go, the more severe this negative energy budget becomes (although no place, not even the North Pole, has a complete or permanent negative energy balance). Even though it is influenced by ambient temperature, cloud cover, and other weather processes, over the long term this negative energy budget is equally distributed on both sides of the winter solstice. At 40° of latitude (e.g., Philadelphia or Denver), the deficit on average lasts only a few days; at 70° of latitude (e.g., Point Barrow, Alaska; Inuvik, NWT), it lasts an average of six months.

Plants of the Boreal Forest

The boreal forest is predominantly, but certainly not exclusively, made up of evergreen, needle-bearing trees. Given the short growing season in the North, plants that carry over their photosynthesis equipment from one year to the next have an advantage. Whenever the weather is warm enough, conifers and other evergreen plants can carry on photosynthesis, using solar energy to transform carbon dioxide and water into life-sustaining glucose. Evergreen trees and herbs thus get a jump start on the growing season.

There may be a second factor causing conifers to dominate the boreal forest. E. C. Pielou, the Canadian naturalist, believes that the domination of conifers in the taiga has as much to do with infertile soils as with the short growing season and the harsh northern climate. She points out that the soils of the North are immature: The continental ice sheets melted away from these lands only about 6,000 to 9,000 years ago, so there has not been much time for these soils to acquire organic material, and the organic material that has built up is often lost in one of the major forest fires that typically occur every century or two in many taiga areas.

Other reasons for the nutrient-poor taiga soils relate to the conditions in a conifer forest. Conifer needles, especially those from black spruce, are acidic, and the acids that drip from them during rainstorms slow the rate of decomposition in the soil and leach nutrients and minerals deeper into the soil and out of many plants' reach. In addition, in a closed-canopy stand of conifers, shade exists year-round. The canopy also intercepts much of the snowfall, which would otherwise form an insulating cover of snow on the ground. Permafrost then forms more readily, further reducing the temperature of the soil. As a result, soils under coniferous stands are cold and decomposition is slow. The next time you are standing in a mature spruce stand in the southern boreal forest, reach down and pull back the duff and litter and dig into the soil. You will probably uncover a light gray soil, about the color and consistency of ashes from a wood-burning fireplace. These nutrient-poor soils are the spodosols (*podzols,* in older terminology) of the boreal forest, and they are about as nutrient poor as they look.

Most plants in the boreal forest use teamwork to wrest nutrients from boreal soils. Conifers, like most higher plants, do not have fine roots to absorb nutrients; they have mycorrhizae—interweavings of fungal mycelium and root tissue. These mycorrhizae constitute an intimate and ancient mutualism. The fungus supplies the higher plant with scarce nutrients and, during drought conditions, water. In turn, the higher plant supplies the fungus with soluble carbohydrates. Most mosses, ferns, lycopods, gymnosperms, and angiosperms of the taiga (as well as other terrestrial biomes) are mycorrhiza formers. (Plants from a few taxonomic families like the Cruciferae [the mustard family] are very much the exception in not forming mycorrhizae. They absorb nutrients and water unassisted through their finely textured roots.)

Nutrient-poor soils tend to favor conifers because many conifers are accumulator species. That is, the minerals—especially calcium, nitrogen, phosphorus, and potassium—that their roots in partnership with mycorrhizal fungi wrest from these nutrient-poor soils are hoarded. Large portions of these accumulated nutrients are not recycled until the trees die and slowly decompose. In fact, across many taiga regions dead plant material is not broken down until forest fires consume the forest.

Pielou observes that, compared with the hardwood trees that dominate the forests of the East, the conifers of the taiga lead frugal lives. They are not as efficient at photosynthesis, yet they don't need to be, because they don't have to grow a full set of leaves each spring. Conifers live on a waste not, want not strategy. Hardwoods, by shedding their leaves each year, live on an easy come, easy go strategy. Their parsimonious lifestyle makes conifers far better able to cope with the short northern growing season and to prosper on the immature, poor soils of the North.

One adaptation vitally important to a tree or shrub growing in the North is cold hardiness. To survive the winter, trees and shrubs do not die back to ground level or go into subterranean retreat the way herbs and grasses do. How do trees and shrubs survive the −40° and colder temperatures regularly encountered during a boreal winter? (Note that −40°F is also −40°C.) Trees are adapted to escape frost injury in one of two very different ways: some do it by supercooling; the remainder do it by extracellular freezing. Supercooling, however, protects trees and shrubs only down to temperatures of −40°, while extracellular freezing is effective well below that temperature. Therefore, as Pielou points out, we have two types of trees: hardy and very hardy. The tree species found in the taiga are all very hardy and use the extracellular freezing method of surviving these subarctic temperatures.

Hardy trees such as the maples, oaks, and other hardwood trees of the eastern deciduous forest allow the liquid in living cells to supercool. More precisely, as the temperature drops the liquids within their cells remain liquids even below 32°F (0°C) because there are no minute particles or rough surfaces on the inside surfaces of the cell walls to act as nuclei around which ice crystals can form. However, if the temperature drops below −40°, the liquid inside of living cells freezes with or without these seed crystals, and the cells of the tree are killed when ice crystals form and disrupt the structure of the cell's proteins and microscopic organelles.

The very hardy trees of the boreal forest have a different way of surviving the cold. In very hardy trees, liquids inside the cells are squeezed out through the cell membrane and freeze in the many small empty spaces within the living tissues of the tree. Since they are not formed inside the cells, these ice crystals do no damage to the tree. A greater portion of conifers as compared to hardwoods use this extracellular freezing technique. Jack pine *(Pinus banksiana)*, lodgepole pine, tamarack *(Larix laricina)*, white and black spruce, and balsam fir *(Abies balsamifera)* are all very hardy trees. Paper birch, trembling aspen, and balsam poplar are broadleaf trees that specialize in this technique. A number of willows and alders are also capable of surviving the deep cold using extracellular freezing. Not surprisingly, these are the main tree and shrub species in the taiga of North America.

Treelines of the Boreal Forest

Much of the variation in the taiga is along its northern edge, where the boreal forest gives way to the Arctic tundra. This forest tundra (the transition zone between the boreal forest and the Arctic tundra) varies in breadth from region to region. It is wide across much of northern Canada, but it occurs in discontinuous patches across Alaska. In central Siberia, the forest tundra is especially wide, and Russian ecologists often treat it as a separate terrestrial biome, not as a transition zone between two biomes. The forest tundra is economically important to Russia, Finland, and Scandinavia, where it is winter range for the domesticated reindeer of the Chukchi, Sami, and other northern people.

The forest tundra ecotone takes on a number of forms. Across much of northern Canada, it is a lichen woodland—an open, parklike forest with stunted black spruce trees growing every ten yards (9 m) or so. These trees are surrounded by a thick carpet of ground lichen (e.g., *Cladonia, Cladina,* and *Thamnolia* species), with a few scattered ericaceous (heather) shrubs. In other areas the forest tundra grows as isolated clumps of black spruce surrounded by large stretches of tundra. In still other areas the forest tundra is made up of extensive stands of willow and dwarf birch, so that the treeline is actually covered by shrubs.

What makes the forest tundra finally give itself over to open tundra? What causes the northern treeline? These questions may sound simple, but actually they are not. For example, the whole concept of a northern treeline is useful only in certain areas. Consider the Mackenzie Mountains, on the border between the Northwest Territories and the Yukon Territory. Here it is difficult to differentiate between a northern treeline and an Alpine (i.e., elevation-related) treeline. In this area, they meld. Or consider northern Norway, where three types of treeline are interwoven: a northern treeline, an Alpine treeline, and a treeline induced by the effects of a cool maritime climate. Consequently, a northern treeline is clearly developed only in certain areas—for example, in Nunavut or the Northwest Territories—but in these areas it is a distinct feature of the landscape and has intrigued naturalists for over a century. Reasonably accurate maps of the northern treeline in these territories have existed since 1845, when the great French naturalist Louis Agassiz published his study.

But what causes a distinct treeline? Why in Nunavut or the Northwest Territories does black spruce finally surrender the land to, as Ernest Thompson Seton calls it, the "great arctic prairie"? The boreal ecologist J. A. Larsen mentions several fascinating theories. The German climatologist W. Köppen suggested in 1936 that the

Northern reindeer lichen *(Cladina stellaris)*, **a circumpolar species that often grows in thick mats on the taiga's forest floor, is an important winter food for caribou and reindeer.**

northern treeline occurs along a line that corresponds with the northern limit of the 50°F (10°C) isotherm for the warmest month. It is an interesting hypothesis and a pretty good fit. More recently, O. Nordenskjold, of Norway, improved the correlation by taking into account not only the mean temperature of the warmest month but also the mean temperature of the coldest month.

R. J. Reed built on this work but suggests an even more interesting theory. His hypothesis grew out of his research on the average summer positions of three air masses: the Arctic, the continental, and the Pacific. An air mass is a region of atmosphere extending horizontally for hundreds or thousands of miles and often extending as far up as the stratosphere. An air mass is characterized by relatively homogeneous atmospheric conditions. Typically, the Arctic air mass migrates south in the wintertime and is the main cause of the frigid boreal winters. During summer, it recedes northward, and the taiga is covered by the warmer continental and Pacific air masses, giving the region its brief warm summer. The belts of converging air where these three air masses meet are understandably regions of unstable weather, and when these frontal zones, or storm tracks, are passing back and forth overhead in the spring and fall, the climate of the boreal forest is typically storm ridden.

Reed carefully documented the average summer position of the border between these three giant air masses. From this research, he shows that the northern treeline coincides with the average summer position of the Arctic air mass. Boreal trees can grow this far north, but at this point the climate cools sufficiently so that the tundra prevails. Furthermore, Reed went on to show that in areas where the boundary between these air masses occupies roughly the same position summer after summer, such as in lands west of Hudson Bay, the northern treeline is distinct and clearly developed. In Labrador and northern Quebec, where the summer positions of the air masses are more variable, the northern treeline is not nearly as clearly delineated.

R. A. Bryson, employing air mass frequency analysis, took Reed's hypothesis even further. He agreed that the northern treeline is determined by the average summer position of the Arctic air mass meeting the Pacific and continental air masses. However, he went on to demonstrate that the southern boundary of the boreal forest is determined by the average winter position of these three air masses. According to this school of thought, the winter and summer positions of these three giant air masses determine the northern and southern boundaries of the taiga. Still missing from these theories is something that relates air mass movement to tree physiology. The southern limit of the boreal forest may be delineated by the northernmost areas where hardy trees and shrubs exist. North of there, due to the cold winter temperatures of the Arctic air mass, only very hardy trees and shrubs survive. However, we don't have all the answers. Why can't black spruce cope with the relatively cool summer air under the Arctic air mass? Although we are beginning to understand the dynamics of the boreal forest, a complete understanding of the treeline remains elusive.

Mammals and Birds of the Boreal Forest

Species diversity falls off markedly as one moves north. The forests of Costa Rica in Central America support 163 mammalian species. Mexico supports 150. Historically,

California supported 115. Much of the taiga of Canada supports only 45. Move north to treeline, and the number is reduced to 35. Move on to the Queen Elizabeth Islands in the high Arctic, and only 8 species of terrestrial mammals are present. This same pattern is also found in the species diversity of insects, birds, and plants. Species diversity is reduced the farther north one goes.

Boreal mammals tend to be habitat generalists. Bill Pruitt states that the hallmarks of boreal mammals are their flexibility and adaptability to the changing conditions of the North. The wolf *(Canis lupus)*, cougar *(Felis concolor)*, red fox *(Vulpes vulpes)*, beaver *(Castor canadensis)*, muskrat *(Ondatra zibethicus)*, and deer mouse *(Peromyscus maniculatus)* are all found in the taiga of Canada, and these species have some of the widest distributions of any North American mammal.

Other patterns among mammals also become evident. Bergmann's rule states that there is an overall increase in body size as one moves north. Allen's rule observes that there is a tendency toward a decrease in the size of the extremities with increasing cold. A comparison of the length of limbs, ears, muzzle, and tail of an Arctic fox *(Alopex lagopus)* and a southern kit fox *(Vulpes macrotis)* illustrates Allen's rule quite well. While these rules, which were put forth nearly a century ago, give good overall guidance, there are plenty of exceptions. For example, Peary caribou from the Arctic Islands is the northernmost caribou, but it is the smallest, rather than the largest, subspecies of *Rangifer tarandus*. Grizzly bears and moose reach their largest sizes in southern Alaska rather than in areas north of there.

S. S. Shvarts, the Russian mammalogist, claims that mammals most adapted to the North commonly show two reproductive traits. One trait is that they reproduce early in the year. For example, lemmings can reproduce during winter if snow cover is good and food under the snow is plentiful. Snowshoe hares *(Lepus americanus)*, red backed voles *(Clethrionomys rutilus)*, and red foxes are other species that give birth to offspring in late winter if conditions are favorable. The second trait is that these northern species have many offspring during favorable years and few or no offspring during unfavorable years. Both traits contribute to the population cycle observed in lemmings, snowshoe hares, lynx, and other mammals in northern ecosystems (see Chapter 8).

Mammals of the boreal forest include red squirrels, nocturnal flying squirrels, woodchucks, porcupines, muskrat, beavers, snowshoe hares, moose, black bears (and in western mountainous regions, grizzly bears), caribou, deer, as well as mice, voles, and ground squirrels. Mountains of the boreal region support endemic species such as bighorn, Stone, and Dall's sheep; mountain goats; and mountain caribou—but these species spend a good deal of their life in cliffs, Alpine tundra, or other non-forested areas. Carnivores include red foxes, coyotes, gray wolves, lynx, marten, fisher, several weasel species, mink, river otter, and wolverine. Cougars are also known to inhabit regions of the boreal forest, but the eastern population of cougar is exceedingly rare and has been declared an endangered subspecies.

The Canadian taiga is an amazing environment during spring, when millions of birds wing northward to their summer breeding ground. After the quiet of winter, the taiga erupts with birdsong. The concert begins in late February with the calls of chickadees and the nocturnes of owls. The crescendo of territorial singing peaks during June and then disappears during July, but the coda of bird calls carries on into the fall until the last few scattered voices of ravens and gray jays and the protests

The beaver, one of the largest rodents in the world, shows complex social behavior and an impressive ability to alter its environment through dam building and the harvesting of trees.

Red fox kits stay close to their den, waiting for one of their parents to return with food.

of nuthatches carry us back into winter. Spring is the most exciting season for birding. Each year when I hear for the first time the gentle, haunting whisper of the wings of white pelicans, I know the season has begun anew. During April and early May, each day brings new bird species returning, species that haven't been seen since the previous autumn. Some of the shorebirds, such as the upland sandpiper *(Bartramia longicauda),* have flown to the pampas of southern Argentina and back again during the intervening months. Arctic terns *(Sterna paradisaea)* have migrated to the southern tip of South America or to Antarctica during this time.

In my first summer of fieldwork in the boreal forest, I remember feeling that these northern woods belonged to the birds. It is not like the spectacular but quickly passing migration season on the prairies, where the sky teems with bird life for a few weeks and then becomes relatively quiet again. Instead, in the taiga during the entire spring and summer the whole forest is full of flickering wings. Birds seem to be at home in every part of the boreal forest. Certainly the mosquitoes, black flies, and other insects, which provide a high-protein diet for many of the warblers and other passerines of the taiga, are a large part of why these birds migrate, some nearly halfway around the globe, to breed and raise their offspring in the taiga during its short but fertile summer.

The number of avian species in the boreal forest changes dramatically with the season. During winter the number of species dwindles to 20 or 30. Near treeline the reduction is more dramatic. Churchill, Manitoba, frequently reports only 2 bird species on its Audubon Christmas Bird Counts: 1 black (raven) and 1 white (willow ptarmigan). During summer, on the other hand, the avian community swells to more than 300 bird species.

In fact, in the narrow band of mixed conifer and aspen woods along the southern edge of the boreal forest in Manitoba, Saskatchewan, and Alberta, there are more species of breeding birds than in any other region north of Mexico. This area of southern boreal forest grades into the aspen parkland, and during spring and summer it is a birder's paradise. These breeding grounds are protected in parks such as Riding Mountain National Park, Duck Mountain Provincial Park, and Spruce Woods Provincial Park in southwestern Manitoba; Prince Albert National Park, Narrow Hills Provincial Park, and Greenwater Lake Provincial Park in central Saskatchewan; and Elk Island National Park in central Alberta.

Then there are the irregular migrations. Periodically, northern seed-eating birds or certain raptors that live in the boreal forest year-round suddenly migrate south. These invasions are known as irruptions and occur as dramatic but irregular events due to food shortages in the taiga—a failure of conifer trees to produce an abundance of cones. Irruptions are seen in Bohemian and cedar waxwings, pine and evening grosbeaks, black-capped and boreal chickadees, red-breasted nuthatches, pine siskins, and common and hoary redpolls. Red- and white-winged crossbills are particularly known for their dramatic southern irruptions.

Paul Ehrlich believes that events unfold something like this. A year with a good cone crop leads to higher populations of seed-eating birds. Such a year is often followed by a year with poor cone crops, and the scarce food resources during autumn cause large numbers of seed-eating birds to migrate southward. Many other factors, such as insect abundance during the breeding season, can influence these patterns. However, the Audubon Christmas Bird Count records, which have been collected now for more than a century, document irruptions of these northern birds in areas south of the boreal forest on an irregular basis.

Certain species of hawks and owls from the taiga and tundra also show irruptive migratory patterns. The Manitoba wildlife biologist Robert Nero points out that these raptors feed on hares and small burrowing mammals over widespread northern areas. Rough-legged hawks and northern goshawks as well as snowy, great-horned, short-eared, and great gray owls irrupt periodically in areas south of the boreal forest. Two main population cycles in northern small mammals appear to set off these migratory movements: a four-year cycle among tundra lemmings and a ten-year cycle among the snowshoe hares of the boreal forest (see Chapter 8). Population peaks among these lagomorphs and rodents support denser raptor populations. However, when these prey populations dramatically decline, the raptors respond by migrating, many of them southward. Rough-legged hawks and snowy owls are sighted during winter on the Canadian prairies with approximately a four-year periodicity. Northern goshawks and great gray owls appear in forested areas south of the taiga with approximately a ten-year periodicity.

Birds of the boreal forest can be classified into three groups: year-round residents, short-distance migrants, and long-distance migrants. Year-round residents include northern goshawk; spruce and ruffed grouse; raven; gray and blue jays; red- and white-breasted nuthatches; downy, hairy, three-toed, black-backed, and pileated woodpeckers; black-capped and boreal chickadees; golden-crowned kinglet; white-winged and red crossbills; evening and pine grosbeaks; and great horned, great gray, boreal, barred, and northern hawk owls. Al Smith, the Saskatoon bird biologist, states that these species are entirely dependent on how well we manage the boreal forest; they are the resident bird species, which cope with the long boreal winter.

Short-distance migrants winter mainly in the United States. These birds include bald and golden eagle; great blue heron; common goldeneye; hawks such as northern harrier, sharp-shinned hawk, and American kestrel; shorebirds such as greater and lesser yellowlegs; and passerines such as belted kingfisher, yellow-bellied sapsucker, northern flicker, eastern phoebe, tree swallow, American crow, winter wren, hermit thrush, American robin, orange-crowned and yellow-rumped warbler, dark-eyed junco, and savannah, white-throated, and fox sparrows. As a group, these birds seem to be less affected than others by environmental changes, and on the whole these species are faring not too badly, at least for now.

The long-distance migrants, or neotropical migrants, are birds that winter in the tropical environments of Mexico, Central America, or South America. Examples of these species include osprey; peregrine falcon; Bonaparte's gull; black-billed cuckoo; common nighthawk; ruby-throated hummingbird; western and eastern wood peewees; alder and least flycatchers; eastern kingbird; purple martin; bank, violet-green, and cliff swallows; house wren; veery; Swainson's thrush; warbling and red-eyed vireos; and more than twenty warbler species, including Tennessee, yellow, chestnut-sided, Cape May, palm, and bay-breasted blackpoll; American redstart; ovenbird; northern water thrush; scarlet and western tanagers; rose-breasted grosbeaks; and sparrows such as Lincoln, chipping, and clay-colored. Many of these species appear particularly sensitive to forest fragmentation on their breeding grounds (including deforestation, seismic lines, oil and gas pipelines, and roads). They are also affected by extensive deforestation on their wintering grounds. There is concern for a number of these species. International cooperation in the conservation of these neotropical migrants is badly needed.

In Canada, endangered, threatened, and vulnerable species of wildlife and plants are formally determined on a national basis by the Committee on the Status of Endangered Wildlife in Canada (COSEWIC). This committee defines endangered species as a species facing imminent extinction or, at least, extirpation from its range in Canada and lists the following mammals and birds from the taiga: cougar in Ontario and east; marten in Newfoundland; wolverine in Quebec and east; whooping crane; harlequin duck in Quebec and east; peregrine falcon (*anatum* subspecies), and Henslow's sparrow in Ontario. Threatened mammals and birds (species likely to become endangered if limiting factors are not reversed) inhabiting the taiga are wood bison, woodland caribou in the Gaspé peninsula, in Quebec, and loggerhead shrike in the Prairie Provinces. Vulnerable mammals and birds (species of special concern because of characteristics that make it particularly sensitive to human activities or natural events) that inhabit the taiga are grizzly bear, woodland caribou in Ontario and west, Gaspé shrew in Quebec, wolverine in Ontario and west, short-eared owl, and Caspian tern.

Special conservation programs are under way or are being planned to protect or restore a number of these wildlife populations. Many biologists and wildlife ecologists have devoted their careers to helping these endangered species, and there are some success stories. The wood bison has improved from an endangered to a threatened species, and the white pelican, which was a threatened species, has recovered to the extent that it has been delisted. In addition, progress has been made in the recovery of species such as whooping crane, peregrine falcon, and grizzly bear. While these results are encouraging, it is obvious from the COSEWIC lists that much work remains to be done.

Global Warming and the Boreal Forest

Global climate change is one of the greatest threats to the boreal forest. To understand why warming the frigid taiga threatens its very existence, we need to focus on certain processes taking place in the Earth's atmosphere.

The atmosphere that surrounds the Earth makes life on the surface of this planet possible. During the early Paleozoic era (600 million to 500 million years before present), living organisms existed mainly in the oceans because the atmosphere of the Earth had not developed sufficiently to protect them from the ultraviolet radiation of the Sun. For 100 million years, life was nearly impossible outside the protective shield of water. Today, the ozone layer in the upper atmosphere provides a shield, reducing ultraviolet radiation to tolerable levels, but the thinning of this ozone layer by chlorofluorocarbons (CFCs) and other pollutants is of major concern.

In total, 99 percent of the Earth's atmosphere consists of gases made of nitrogen (N_2) and oxygen (O_2). Almost all living organisms use oxygen to carry out respiration—a metabolic process that gives living organisms the energy to stay alive, while giving off carbon dioxide as one of its waste products. Chemically and physically, O_2 and N_2 have little effect on the flow of heat through the atmosphere. However, other gases, called greenhouse gases—including carbon dioxide (CO_2), methane (CH_4), nitrous oxide (NO_x), and water vapor (H_2O)—make up the remaining 1 percent of the atmosphere, and these gases greatly affect the flow of heat radiating from the surface of the Earth and flowing through the atmosphere.

Like glass panels in a greenhouse, these gases allow sunlight to pass through them unaffected, but when the Sun's energy is absorbed by the land or water and reflected or radiated back into the atmosphere as heat, the longer wavelengths of this heat radiation are partially absorbed by these greenhouse gases. This is a good thing, because the process raises the temperature at the surface of the Earth 60°F (33°C), making most of this planet inhabitable by living organisms. However, because greenhouse gases make up only a small percentage of Earth's atmosphere, it is possible for the emissions from human activities to change significantly the concentration of these gases in the atmosphere. By raising their concentration, human activities inadvertently cause the climate of the Earth to warm.

Analyzing the CO_2 concentration of ice cores taken from the polar ice sheets, scientists have shown that for the past 1,000 years carbon dioxide in the atmosphere has varied little—by only a few percentage points one way or the other. However, from about 1850—at the start of the industrial revolution—to the present day, CO_2 in the atmosphere has increased from 280 to 360 parts per million, or approximately 30 percent. Environment Canada's forecast for the next century predicts that industrial activities, consumption of fossil fuels, increased transportation, and other human activities will double or perhaps triple the amount of CO_2 in the atmosphere. Methane and nitrous oxides will also be increased by these activities. What will be the environmental, social, and economic impacts of increasing these greenhouse gases during such a short period of time? In particular, how will global warming affect the boreal forest?

The temperature at the surface of the Earth varies season by season, year to year, and region by region. However, when climatologists make a complex calculation—the average temperature for the entire surface of the Earth on an annual basis—many of these variations are averaged out. Most scientists studying climate change agree that this average temperature for the surface of the Earth has increased by 1°F (0.5°C) during the past 100 years. What's the big deal about a single degree? Well, it does represent a big change. An increase of a degree in this average temperature exceeds any documented global changes for temperature during the past 600 years. Furthermore, the change over the past 15,000 years, from the last great Ice Age to the present, amounts to a warming of 9°F (5°C) in this overall average temperature at the surface of the Earth. That change transformed the landscapes of Canada from large continental ice sheets, several kilometers thick, to the mosaic of productive ecosystems present in Canada today. Over the past century and a half, human activities have warmed the Earth 10 percent of that amount, with the prospects of warming doubling or tripling during the next century. What changes will that bring?

Due to trade wind patterns and decade-long phenomena such as the Arctic oscillation, climate warming is much more severe in northern regions than for the Earth as a whole. As an example, let's look at Alaska, using the analysis produced by Gunter Weller and his colleagues. Alaska has experienced the largest regional warming of any state in the United States, with a rise in average annual temperature of 5°F (3°C) since the 1960s and of 8°F (4.5°C) during winter. Much of this warming occurred during the late 1970s, with melting of glaciers, thawing of permafrost, and reduction of sea ice. Precipitation increased roughly 30 percent between 1968 and 1990. In addition, the past two decades have shown two of the most intense El Niño oscillations on record, further affecting climate change in Alaska.

Perhaps it is not all our fault. Climates do change naturally, and we can ask: To what extent could this global warming be due to natural causes? In fact, atmospheric scientists have concluded that a portion of the change in temperatures over the past century is indeed due to natural causes such as increased sunlight intensity. These experts estimate that 50 percent of the temperature change observed during the past century and 30 percent of that during the past three decades can be attributed to increased solar radiation. But the remainder appears to be a result of human industrial activities. Furthermore, if these human activities double or triple the amount of greenhouse gases in the atmosphere during the next century, global warming can be expected to accelerate. The average temperature for Canada and Alaska is expected to increase 6°F (3°C) by 2030 and as much as 18°F (10°C) by 2100. Currently, the global climate is warming ten times faster than it did at the end of the last Ice Age, and its pace is expected to accelerate.

The implications of continued global warming for the North are serious. A study carried out by Jay Malcolm for the World Wide Fund for Nature predicts that 46 percent of the natural habitat in Canada will change so dramatically that it will be unable to support the species of wildlife and plants now living in these areas. Habitat change is predicted to be even more dramatic for certain of the provinces and territories: Alberta is expected to lose 56 percent of its natural habitat; Ontario, 61 percent; Newfoundland, 64 percent; and the Yukon, 64 percent. New habitats will take their place, but the fact that the climate is changing ten times faster than at the end of the last Ice Age may mean that these new habitats will be dominated by invasive species of plants and nonnative species of animals.

While Alaska is getting wetter, other regions are getting dryer. The relationship between average temperature and evaporation of water from the surface of the land and lakes is important. At the Experimental Lakes Area in the boreal forest of northern Ontario during the past decade, an increase in the average temperature during the snow-free season from 57° to 60°F (14° to 16°C) caused a 30 percent increase of evaporation from the land and lakes. If only a few degrees of warming caused almost a third more water to evaporate from the land and lakes, what changes will 20° bring?

As the climate changes, much of Canada's taiga, especially in the Prairie Provinces, is expected to experience drought. One scientific prediction is a 200 percent increase in the amount of forest burned per decade in the boreal forest across North America. Increased drought in certain areas may lead to large insect outbreaks similar to the infestation of spruce bark beetle *(Dendroctonus rufipennis)* now occurring from the Kenai Peninsula in southern Alaska across the southern Yukon Territory. By 1999, 2.3 million acres (930,000 ha) of mature spruce forests in Alaska have died as a result of these infestations. The presence of drought-stressed trees less able to resist spruce bark beetle attacks and a series of milder winters resulting in reduced beetle mortality have led some scientists to attribute these beetle infestations to global climate change.

In northern Alaska, the geologists A. H. Lachenbruch and B. V. Marshall have made measurements in deep boreholes drilled into continuous permafrost. Their research shows warming of up to 7°F (4°C) over the last century. Discontinuous permafrost throughout Alaska has warmed, and some of it is currently thawing from both the top and the bottom. As Gunter Weller points out, the disappearance of this permanently frozen ground is expected to cause landslides, siltation of rivers and lakes, decreased success of spawning fish, slumping of ice-rich landscapes, and massive erosion along the shorelines of rivers, lakes, and oceans. At some points along the Arctic coast, land has receded 1,500 feet (500 m) since the 1950s, and important archaeological sites have been lost. Changes such as these threaten many human communities in coastal locations throughout the North. In addition, in Siberia, Russian engineers are concerned about the safety of apartment buildings and pipelines built on permafrost terrain. Many of these structures have already failed. In the entire discontinuous permafrost zone of the Arctic, roads, buildings, airfields, and pipelines are under threat.

Warming trends in the Arctic are also believed to have caused the decrease of some 155,000 square miles (401,000 km^2) of sea ice over the past decade. This is an area nearly the size of California. Several global climate change models predict by 2100 a near-total melting of the Arctic pack ice during summer.

Changes to boreal lakes and rivers are of equal concern. Thousands of boreal wetlands are expected to dry up and disappear. As these wetlands and small lakes dry up, waterfowl production will drop across the Prairie Provinces and in such areas as the Peace-Athabasca delta in northern Alberta and the Old Crow flats in the northern Yukon. Certain fish populations are expected to show drastic declines. Some predictions indicate that oceans will become too warm for any salmon species to survive.

Global warming will change the basic chemistry of boreal lakes, making many of them more susceptible to the effects of acid rain. It will also alter their fish populations from cold-water species such as lake trout *(Salvelinus namycush),* grayling *(Thymallus arcticus),* and dolly varden *(Salvelinus malma)* to warm-water species such as northern pike *(Esox lucius)* and walleye *(Stizostedion vitreum).* As the climate warms, the invasion of boreal lakes and rivers by nonnative species of fish, plants, invertebrates, and diseases is one of the greatest threats to their ecological integrity. The invasion of the Great Lakes by the zebra mussel is one such example (see Chapter 10).

David Schindler, one of Canada's leading aquatic ecologists, stresses that the effects of global climate change will not occur in isolation. They will interact with other factors such as the overexploitation of fish populations, dam building, water-diversion schemes, habitat destruction, and the spread of nonnative species and pollution. From the interactions of all of these environmental stresses, he predicts that the native freshwater fish populations of Canada will largely be destroyed by 2100. As many lakes and rivers dry up, he further believes that great pressure will be exerted on Canadians to share their water resources with the rest of the world. For example, the Oglala aquifer, which provides water for much of the west-central region of the United States, is being exploited eight times faster than its waters are being renewed by natural sources. These conditions will worsen as the climate warms. Demand for Canadian freshwater will increase. Schindler wonders if proposed disruptive water schemes will be implemented. For example, the GRAND canal scheme proposes to dam James Bay, making it into the largest freshwater reservoir in the world. The project would then build a massive series of canals, locks, power plants, and dams to divert the water of James Bay to Georgian Bay on Lake Huron, where it would be flushed through the Great Lakes to feed pipelines extending all the way to the southwestern United States.

These are some of the documented and predicted impacts of global climate change not only on the boreal forest but also on the rest of North America. With these kinds of impact hanging in the balance, what is our society doing about this global problem? At the present time, the answer is, very little. The strongest response can be found in the communities of the North (see, for example, www.taiga.net/nce). It is impressive to see small communities across the North taking action in anticipation of the effects of global climate change, even though on a national scale very little has been accomplished. Northern communities are changing their construction practices on the conviction that permafrost will no longer be stable. They are locating new schools and nursing stations sometimes several miles from their present community so as to build them on sites they know will be stable in the future. In addition, roads are being rerouted and landing strips are being rebuilt on safer ground.

These are adaptations to the problem, but precious little is being done to address the causes. One of the most hopeful developments is the Kyoto Protocol on Climate Change, which was drafted at an international convention of representatives from 166 countries in Kyoto, Japan, in December 1997. Canada's pledge was typical: The federal government promised to bring Canadian emissions of CO_2 to a level 6 percent below its 1990 level by 2010. The Kyoto Protocol has been heavily criticized by North American automobile and petroleum industries. Since 1997 the CO_2 emissions of countries such as Canada, the United States, Australia, and Japan have actually increased, and it looks as if their 2010 commitments will not be met. In November 2000 a conference was held to assess progress in meeting the Kyoto commitments: While European countries have reduced their CO_2 emissions, Canada, the United States, Australia, and Japan were heavily criticized for their lack of progress. During the spring of 2001, President George W. Bush announced that the United States would not meet its Kyoto commitments, and Prime Minister Jean Chrétien has indicated that Canada is reconsidering its commitments. Activists Maude Barlow and Tony Clarke state that the Kyoto commitments represent only 10 percent of what needs to be done to address the problems of global climate change. They observe that it is disappointing that some of the richest countries in the world won't take even the preliminary step that Kyoto represents toward solving this global crisis.

General References

Agassiz, L. 1845. Essai sur la geographie des animaux. Rev. *Suisse er Chron. Litt.* 8:441–52, 538–51.

Barlow, M., and T. Clarke. 2001. *Global Showdown: How the New Activists Are Fighting Global Corporate Rule*. Toronto: Stoddart Publishing.

Ehrlich, P. R., D. S. Dobkin, and D. Wheye. 1988. *The Birder's Handbook*. New York: Simon and Schuster.

Environment Canada. 1998. *Frequently Asked Questions about the Science of Climate Change*. Report 98-2. Downsview, Ont.: CO_2/Climate Reports.

Lachenbruch, A. H., and B. V. Marshall. 1986. Changing Climate: Geothermal Evidence from Permafrost in the Alaskan Arctic. *Science* 234:689–96.

Larsen, J. A. 1974. Ecology of the Northern Continental Forest Border. In *Arctic and Alpine Environments*, edited by J. D. Ives and R. G. Barry. London: Methuen.

Lynch, W. 2001. *The Great Northern Kingdom: Life in the Boreal Forest*. Markham, Ont.: Fitzhenry and Whiteside.

Nero, R. W. 1980. *The Great Gray Owl*. Washington, D.C.: Smithsonian Institution Press.

Pielou, E. C. 1988. *The World of Northern Evergreens*. Ithaca, N.Y.: Comstock Publishing Associates of Cornell University Press.

Pruitt, W. O., Jr. 1978. *Boreal Ecology*. Studies in Biology 91 (Institute of Biology). London: Edward Arnold Publishers.

Weller, G., P. A. Anderson, and B. Wang. 1999. *Preparing for a Changing Climate: The Potential Consequences of Climate Variability and Change*. Fairbanks: University of Alaska, Center for Global Change and Arctic System Research.

3
Knobs, Kettles, and Precambrian Corks

In the broadest terms, the ecological regions of Canada run east and west, through the grassland prairies and deciduous forests of the south and the boreal forest and Arctic tundra of the North. The geologic regions of Canada, on the other hand, run north and south, from the Appalachian Mountains on the eastern seaboard to the cordillera in the west. The matrix of these regions yields a checkerboard pattern. This chapter undertakes to understand a good portion of that checkerboard.

More precisely, this chapter reveals something about the geologic regions underlying the Canadian taiga, beginning with the major types of rock formation and how they formed. We especially need to know how the Canadian Shield, which serves as the core of the North American continent, originated. We also look at the other geologic regions that the taiga occupies and examine some of the common landforms created by the Pleistocene Ice Age that are typical of taiga country. With this grasp of geology and geomorphology, we will be better able to understand the ecology of the taiga and the portion of the Canadian checkerboard that it claims as its own.

Classifying Rocks

Geology mainly deals with the study of rocks. Rocks are composed of minerals, which are in turn composed of atoms, or chemical elements. Some minerals, such as gold, silver, iron, and platinum, are each made of just one kind of atom, but most minerals are chemical compounds, combining two or more elements.

Minerals combine in many different ways and form many kinds of rock. The rocks that make up the crust of the Earth can be classified into three large groups: igneous, sedimentary, and metamorphic. Igneous rocks are "made from fire." Sometimes magma (liquid rock) from the hot center of the Earth breaks through the crust and comes oozing out onto the Earth's surface. When this molten rock cools, igneous rock forms. The two most common kinds of igneous rock are granite (magma cools and hardens slowly while entrapped in other rock) and basalt (magma cools and hardens more quickly on the Earth's surface or the ocean floor).

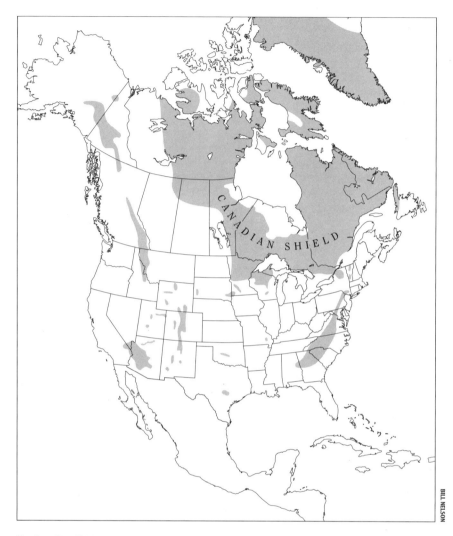

The Canadian Shield

Sedimentary rock forms from the weathering of other rocks. Wind, rain, ice, heat, and cold all wear rocks down, breaking them into fragments. Rain and snow carry these fragments into streams and rivers, which grind them into gravel, sand, and mud, ultimately carrying them into seas and oceans. These sediments are laid down in dense layers in calm water, and the layers of sediments are heated or pressurized into layers of sedimentary rock. For example, sand is compressed into sandstone, and the accumulated shells of certain sea creatures are pressed into limestone. Sedimentary rock forms in layers, and if there has not been disturbance or folding of these layers, it is safe to assume that the layers on top are younger than the layers found beneath.

Metamorphic rock has had its composition and form changed due to the effects of intense heat or pressure acting upon it. Movements of the Earth's crust, or intrusion of molten lava, can provide the intense heat and pressure necessary to make igneous rock or sedimentary rock change into metamorphic rock. For example, granite metamorphoses into gneiss (pronounced *nice*), and limestone metamorphoses into marble.

The Age of the Earth

The Earth must certainly be as old as its oldest rock formations. Some of the oldest rocks found anywhere are rock formations found near Nuuk, in southwestern Greenland. Geologists age rock formations using radioactive isotopes within the rock. These are atoms of a certain element that have an unusual number of neutrons, and thus these atoms have an unstable structure and periodically give off radioactive waves and particles. Gradually, these radioactive isotopes transform into stable atoms of a different, simpler element (the daughter product), and their radioactive decay stops. These unstable isotopes are incorporated as trace elements in igneous rock when it forms, and then they gradually tick down, forming stable, simpler atoms at an overall predictable rate for each element (called a half-life), which varies from a few hundred years for carbon 14 (the number is the atomic mass of the atom) to 4.5 billion years for uranium 238. Consequently, by determining the amount of daughter product found in a rock specimen and then calculating backward, radioactive isotopes can be used as tiny geologic clocks with which to date the origins of rock formations. Dated this way, the rock formations near Nuuk are 3.8 billion years old, give or take a few million years.

Another research approach pushes the age of the Earth back to 4.6 billion years. This technique compares the age of the oldest rock formations found on Earth with the age of meteorites that have collided with the Earth over its long history. Comparison of the radioactive isotopes in these ancient rocks and meteorites suggests that they both formed at about the same time. Furthermore, radiocarbon dating of rock samples brought back from the Moon suggests that it (and perhaps the Sun and most of our solar system) formed approximately 4.6 billion years ago.

Based on patterns in rocks, geologists have created a time scale dividing this 4.6-billion-year (4,600-million-year) time span into four geologic eras: Precambrian, Paleozoic, Mesozoic, and Cenozoic. Each era is subdivided into periods that correspond to major geologic events and major fossil assemblages that trace out the history of the Earth.

The Precambrian Era. This era was by far the longest. It lasted from 4,500 million to 570 million years ago, or 88 percent of the planet's history. During almost all of this time, the Earth was a place hostile to most forms of life. Volcanoes erupted; mountain ranges formed and were worn down by erosion. The most common forms of life were prokaryotes—bacteria and blue-green algae, single-celled organisms that lack a membrane-encased nucleus. The concentration of oxygen and ozone in the atmosphere was low, and ultraviolet radiation at the surface of the land and water was probably lethal, so that life forms existed only at certain depths in lakes and seas. Few of these living organisms possessed shells or hard parts, so that fossils found in Precambrian rock formations are exceedingly rare. Some of the oldest fossils found anywhere in the world have been discovered in the taiga region of Canada. These ancient fossils—stromatolite algae—are preserved in the Precambrian rock around Steep Rock Lake in northwestern Ontario. They resemble petrified cabbage heads. These primitive plants appear to be 2,600 million years old. Steep Rock Lake is approximately 130 miles (210 km) west of Thunder Bay, Ontario, just north of Quetico Provincial Park and the town of Atikokan. These rare fossils are unique heritage resources, and they certainly deserve more protection than they are at present receiving.

The Paleozoic Era. *Paleozoic* means ancient life. This era lasted from 570 million to 245 million years ago. A great variety of hard-shelled arthropods and corals are some of the life forms preserved in fossils found in rock formations that date from this era. The Burgess Shale in southeastern British Columbia formed 520 million years ago and is an important deposit of Paleozoic fossils. Stephen Jay Gould, in his book *Wonderful Life,* writes about the tremendous diversity of phyla preserved there: "The Burgess Shale, in the Canadian Rockies, contains the world's most important fossil fauna." This site is now protected and managed as part of Yoho National Park.

The Mesozoic Era. *Mesozoic* means middle life. This period lasted from 245 million to 66.4 million years ago. Mesozoic oceans teemed with primitive fish, and the land was occupied by dinosaurs. Recent research shows that dinosaurs were much more widespread than previously believed. Dinosaur fossils have been found on every continent. R. A. Gangloff and K. C. May, of the University of Alaska Museum in Fairbanks, document more than a hundred species of dinosaurs in the Coville River area, on the North Slope of Alaska. The Mesozoic era ended with the extinction vortex, which brought the Age of the Dinosaurs to an abrupt end. The Red River valley and, in particular, Dinosaur Provincial Park, in southern Alberta, as well as the Royal Tyrrell Museum, in Drumheller, Alberta, are rich locations for seeing the fossils of dinosaurs and other Mesozoic life forms.

The Cenozoic Era. *Cenozoic* means recent life. This era, which began 66.4 million years ago, includes the present day. Some of the major life forms of this era are flowering plants, insects, warm-blooded mammals, and birds. The last 2 million years of this era have witnessed the Pleistocene Ice Age, during which the northern and southern latitudes of the planet were covered by nearly continuous ice sheets that formed and then melted and reformed again. Canada offers a number of fine museums that interpret the life and times of the Cenozoic era and describe the way the Pleistocene Ice Age sculpted the landscapes of Canada. Among these museums are the National Museum of Natural History in Ottawa (Ontario), the Museum of Man and Nature in Winnipeg (Manitoba), the Museum of Natural History in Regina (Saskatchewan), the British Columbia Provincial Museum in Victoria (British Columbia), Prince of Wales Northern Heritage Center in Yellowknife (Northwest Territories), and the Beringia Centre in Whitehorse (Yukon Territory). Each of these museums is well worth visiting (see Chapter 14).

Drifting Continents and Precambrian Corks

Few things seem as solid and unchanging as the bedrock surface of the Earth. Yet everything is relative. There is clear evidence that the rock formations of Canada, as well as of the rest of the Earth, change constantly over the millennia. In geologic time, mountain ranges are worn down, volcanoes erupt, and layers of sedimentary rock are lifted up and then weather away.

Geologists account for much of this change through the theory of continental drift. According to this theory, the whole continent of North America is in constant motion. Until the 1960s very few geologists believed that the continents could possibly move. Today, most Earth scientists embrace this theory and accept that the more general and important theory of plate tectonics is the best explanation for drifting continents and the changing surface of the Earth. In fact, the last forty years have been a time of revolutionary geologic thought and research.

To understand these related theories, it is best to start with the Precambrian shields, the ancient, granitic rock formations that form the core of each of the seven continents. The core of the North American continent is the Canadian Shield. It is also the geologic region that underlies more than half of the Canadian taiga.

The granitic rocks that form the Precambrian shields are relatively light as far as rocks go. Granites are normally made of silicon, oxygen, potassium, and sodium, as well as other elements. They are light in density compared to basaltic rocks, which are normally made of iron, magnesium, manganese, and other elements. In granitic rocks the elements form minerals of low density, such as quartz, feldspars, and micas. In basaltic rocks the elements form much denser minerals, such as olivine, pyroxene, and metallic oxide. Furthermore, the rate of cooling determines the size of the mineral crystals. Granitic rocks coalesce slowly, cool gradually out of liquid magma, and form slowly, so that their crystals grow large. Basalts cool and solidify more quickly, thus creating small crystals, pressed together in a dense mass. Basaltic rocks typically form the ocean floors. Great masses of granitic rocks form the crust of the continents. Together, basaltic and granitic rocks form the outer crust of the Earth.

Precambrian shields, which are more correctly called Precambrian cratons, form the cores of the seven continents. Cratons are immense blocks of ancient rocks that have remained relatively unaltered for billions of years. Onto these cratons are welded orogenic belts—that is, elongated regions of crust of varying ages and origins that have often been intensely folded during mountain-building processes. The cratons, together with orogenic belts and other fragments that have been cemented on, make up the continents. Thus the continents float like gigantic icebergs suspended in a thinner layer of ever-shifting pack ice. In this analogy, the pack ice is the thinner crust layer of the oceanic floors. The lower density of granitic rock, out of which the continents are mostly made, helps to keep them afloat as they bob along, suspended in the crust of the Earth.

Scientists believe that approximately 350 million years ago all the continents were joined, forming one huge supercontinent, which Alfred Wegener, the main originator of the continental drift theory, named *Pangaea* (a Greek word meaning all the earth). Drawings of this ancient supercontinent show South America nestled up to the coast of Africa and Newfoundland and Quebec snuggled up to Spain and England. About 200 million years ago, Pangaea, because of the movement of gigantic tectonic plates underneath it, began to break apart into seven distinct chunks, creating the plates that contain the continents of today.

In fact, Pangaea may be only the most recent assemblage of the continents into a super continent. The immense crust layer of the Earth is divided into six large plates and about twelve smaller ones. Embedded in seven of these tectonic plates are the continents. The plates float on the huge semimolten layer of the upper mantle that underlies the crust layer and are driven by the intense heat escaping from the center of the Earth. As a result, the plates are in constant motion. Thus the geologic history of the Earth is a history of these plates and continents colliding and separating. As a result, the locations and shapes of not only the oceans but also the continents have been changing constantly over the 4.6 billion years that the oldest parts of the crust of the Earth have existed.

Probably at several times over the history of the Earth the continents have been swept together and welded into a supercontinent. However, this gigantic landmass always contains seeds of its own destruction. A supercontinent acts as a layer of insulating rock over the underlying mantle. Heat builds up under the supercontinent, so that huge columns of magma (called thermal plumbs) rise up and erode the rock structure underlying this supercontinent. The thermal plumbs allow large amounts of magma to intrude into the continental crust, weakening the supercontinent and causing its lifting and breakup. This process, in whole or in part, may have occurred several times during the Earth's history, and our seven continents of today are only a momentary glimpse of an ever-changing kaleidoscope of tectonic fragments.

There have been other planet-shaking events over the past 200 million years. Parts of continents have broken off and smashed into the other continents, relatively minor collisions as far as geologic events go. For example, the area we call India broke loose from Antarctica and Africa, migrated northward, and collided with Asia. The dents and wrinkles from that collision we call the Himalaya Mountains. Volcanic landmasses and oceanic fragments have also moved across the Pacific and collided with the west coast of North America, giving rise to our western mountain ranges.

When the borders of two tectonic plates collide, they usually form a zone of subduction, in which one tectonic plate, usually the one made of the denser basaltic rock, is turned down under the plate of lighter density. If the portion of the second plate consists mainly of a continent, the edge of that plate bobs around in the zone of subduction but is not pulled downward into the mantle of the Earth and remelted into liquid magma. As far as geologists can tell, all of the fragments of Pangaea are still on the move; none have come completely to rest. For example, the Indian tectonic plate is still driving into the main Asian continent. But rest assured, standing on the Canadian Shield our feet are on solid ground—as solid as a cork bobbing around in the eddy of a mountain stream. Nevertheless, the Canadian Shield forms an important part of the North American craton, the core of our continent. This largely granitic core is immense—it underlies most of the North American continent east of the western mountains. Only a third of this Precambrian craton is exposed on the surface of the ground; that portion occurs mostly in Canada, and we call it the Canadian Shield. The Adirondack Mountains of New York State are part of this shield. The exposed bedrock of this Precambrian shield forms most of Quebec, the northern portions of Ontario, Manitoba, and Saskatchewan, and most of Nunavut and the Northwest Territories. It is one of the main geologic regions that supports the Canadian taiga.

If the jewel-like beauty of the Canadian Shield is not enough, if knowing that much of the shield is made of the roots of mountain ranges (mountains that once stood higher than the Rockies) is not enough, then one can ponder this fact: Although the Canadian Shield is composed of some of the world's oldest rock formations, at the same time it is one of the youngest landscapes on Earth. Most of the shield's rock is billions of years old (the oldest formations date back 3.9 billion years); yet over the past 2 billion years much of this Precambrian rock has been reworked, refolded, and remelted to create the metamorphic complexities present on the shield today. The surface features of the shield are much younger, however—the result of Pleistocene glaciations. The last ice sheets retreated from the Canadian Shield only about 6,000 to 9,000 years ago. The surface characteristics of the shield are dominated by the effects of this glaciation and by the subsequent freeze-thaw cycles that mark the taiga year.

The Geologic Regions of the Canadian Taiga

The Canadian Shield occupies more than half the land of the Canadian taiga. However, the taiga also grows on three other geologic regions. Instead of limiting ourselves to these four regions, let's examine all six of Canada's geologic regions.

Describing the geologic regions of Canada is a little like describing a bug by using a microscope. The way the scientist identifies the parts of the bug and describes the structure of each part depends upon what power of the microscope is being used. Different parts and a new layer of structure become evident with every major increase of magnification. Describing the geology of Canada is much the same: if we turn up our geologic microscope, different regions and subregions, with different layers of structure, become evident. In this chapter we work at the lowest power, which gives an admittedly broad-brush description of each region—but one that is appropriate for the scope of this book.

Even in the broadest of terms, Canada is a complex collection of geologic regions: the enormous Precambrian shield core, the Appalachian Mountain region, the Canadian Cordillera, the Innutian Mountains, the Interior Plains and lowlands, and the continental shelves. Each of these regions is subdivided, a classification system that changes with new methods of dating rock formations and new research on plate tectonics and other geologic processes. However, for the purpose of this book we follow the *Canadian Encyclopedia* and classify Canada into the six regions listed. The Canadian taiga occupies the northern portions of four of these regions.

The Canadian Shield. This core of the continent, by far the largest geologic region of Canada, almost encircles Hudson Bay. It occupies approximately half of Canada's 3.74 million square miles (9.97 million km^2). Its rocks, formed during the Precambrian era, are among the oldest on Earth. The long geologic history of the Canadian Shield is complex and obscure. Evidence of the early parts of its history have been largely erased or eroded. On several occasions, the shield's rocks formed high mountain chains, but weathering and erosion wore them down so that large parts of the shield today are simply the roots of these mountains. Rock formations were often turned into metamorphic rock, and deposits of valuable minerals, such as gold, nickel, silver, and lead, have formed in veins running through these metamorphic formations. Starting about 900 million years ago, the shield became stable. To this stable core, all the other geologic regions of North America were later attached.

The Appalachian Mountain Region. Bordering the Atlantic coast of Canada and the United States, the Appalachian mountain belt extends from Alabama to Newfoundland and covers the Atlantic provinces and southeastern Quebec. In early Paleozoic times an ocean trough southeast of the Canadian Shield was filled with sediments. These sediments were eventually compressed into rock and uplifted to form the original Appalachian Mountains. Since then these mountains have been worn down and raised up many times. Some of the sediments from these eroding mountains gradually buried great forests of club mosses, horsetails, and ferns, which dominated the vegetation of this region. These are the origins of the huge coal formations in these mountains today.

The Canadian Cordillera. This region resembles an accordion. Large volumes of sediments were laid down in seas west of the Canadian Shield. The sediment then formed layers of sedimentary rock, which were folded and uplifted to form high mountain ranges. This orogeny (mountain building) happened repeatedly toward the end of the Mesozoic era, and the result is the Canadian Cordillera. Elsewhere, foreign fragments in successive waves collided with and were spliced onto the western coast of the North American continent during the mid and late Mesozoic era. Each of these continent-building accretions was followed by mountain-building episodes and granitic intrusions from the Earth's mantle below. Since the late Mesozoic the Pacific plate has been sliding under the cordillera region, causing earthquakes and volcanoes along the west coast. At the same time, the North American plate continues to migrate westward, making its way toward Asia and overriding the Pacific plate as it goes. As a result, we see a mix of both sedimentary and metamorphic rock mountains, twisted, piled, and tilted into the impressive cordillera.

The Innuitian Mountains. This mountain belt, more than 795 miles (1,280 km) long, stretches across the most northerly of Canada's Arctic Islands. Like the Appalachian Mountains, the rocks of this mountain belt are sedimentary. They are of a variety of ages. They were squeezed and folded into mountains after the Appalachian Mountains had formed. Because of their remote northern location, less mineral exploration and geologic research have been carried out in this region compared to the rest of Canada.

Interior Plains and Lowlands. This is not one continuous geologic region but rather a set of subregions that share a common origin and that are scattered across the country. The region includes the Interior Plains of western Canada, the lowlands of Hudson Bay and Ungava Bay, the lowlands in southern Ontario and along the

St. Lawrence River valley, and the lowlands of the Arctic Islands. On several occasions during the Paleozoic era, the edges of the Canadian Shield were flooded by shallow seas. Before the rise of the cordillera, sedimentary rock, particularly sandstone, were laid down in these shallow seas. When the mountains and surrounding areas were uplifted and the seas retreated, this rock formed the plains of western Canada and of the various lowlands. During the mid-Paleozoic era, before these areas were uplifted, abundant populations of tiny plants and animals lived in the shallow seas. As they died, they were buried in the sediment and turned into the rich oil and gas reserves of Alberta and Saskatchewan. As water evaporated from the huge saltwater pools near these shallow seas, minerals precipitated and were deposited to create the huge potash and salt formations of the prairies—which are commercially mined today.

The Continental Shelves. These shelves are extensions of Canada's landmass under the Pacific, Atlantic, and Arctic Oceans. They form submarine terraces up to 450 miles (725 km) wide and up to 1,000 feet (305 m) below the ocean's surface. The continental shelves are the youngest additions to the North American continent and consist mostly of sedimentary rock. The Arctic and Atlantic shelves developed along stable continental margins building up on ancient ocean floor. The Pacific shelf, however, formed along the edge of the North American plate that is sliding over the oceanic crust of the Pacific plate, and parts of this shelf have broken off and been subducted. It is not hard to understand why the Pacific shelf is not as broad or expansive as the other two. During past periods these shelves were areas of warm, shallow seas, where a rich growth of plants and animals occurred. Remains of this abundant life became the petroleum reserves located off the Atlantic coast as well as the oil and gas deposits found in the Beaufort Sea and other places in the Canadian Arctic.

The two geologic regions outside of the taiga are the Innuitian Mountain region and the continental shelves. On the remaining geologic regions—the Canadian Shield, the Appalachian Mountains, the Canadian Cordillera, and the plains and lowlands—the taiga has adapted itself.

A Land Carved by Ice

To understand the land upon which the taiga of Canada grows, we must also understand the landforms created by the glaciers and the ice sheets that covered Canada for thousands of years. At the height of Pleistocene Ice Age, 97 percent of Canada was covered by glacial ice. During the major and minor glacial ages of the past 7 million years, ice sheets and mountain glaciers expanded and contracted in a cycle of approximately 10,000 years. Among the numerous cycles of the Pleistocene epoch, eight major ice ages can be identified. It usually took many thousands of years for ice caps and glaciers to build up to their maximum size and extent, but sometimes it took only hundreds of years for these immense ice sheets to melt away.

The glaciologist S. E. White points out that there have been other ice ages during the long history of the Earth. The Vendian Ice Age occurred 675 million years ago, toward the close of the Precambrian era. The Ordovician-Silurian Ice Age occurred around 450 million years ago, and the Permian Ice Age marked the close of

Paleozoic times, around 245 million years ago. Then there were no major ice ages until the Pleistocene Ice Age, which began approximately 1.5 million years ago and continued until 10,000 years ago—if indeed it has truly ended.

Why did these ice ages occur? Most glaciologists admit that we do not understand the causes. Some of the theories being researched include changes in solar flares or sunspot numbers over time, which influenced the climate of the Earth; periodic perturbations of the Earth's orbit around the Sun, with dramatic effects on the Earth's climate; greater or lesser amounts of carbon dioxide, ozone, or volcanic dust in the Earth's atmosphere, reducing the amount of solar radiation received at the surface of the Earth; the drifting of continents toward polar positions; and changes in the atmosphere-to-ocean circulation patterns, affecting heat exchanges between areas of high and low latitudes. As interesting as this topic is, here we limit ourselves to understanding how the Pleistocene glaciers and ice sheets created many of the landforms of Canada.

By definition, a glacier is any large mass of perennial ice that forms on land through the compaction and recrystallization of snow into ice and that forms a mass of ice great enough to flow outward as a result of its own weight. The term *ice sheet* is commonly applied to a glacier occupying an extensive tract of relatively level land and exhibiting flow from the center outward. Glaciers and ice sheets occur when the snowfall in winter exceeds the melting in summer, conditions that at present prevail only in high mountain regions and polar regions. Glaciers occupy only about 11 percent of the Earth's land surface. However, they still hold roughly three-quarters of the freshwater reserves of the Earth. During the height of the Pleistocene Ice Age, glaciers and ice sheets covered 30 percent of the Earth's land surface, and sea levels around the world were considerably lower.

Approximately 99 percent of glacial ice is concentrated in Antarctica and Greenland; the rest is in the more than 200,000 glaciers in mountains from the equator to the polar regions. If all of this glacial ice were to melt, the resulting rise in sea level would be between 200 and 330 feet (60 and 100 m)— enough to submerge every major coastal city. The melting of glacial ice is another concern that scientists are researching connected to the problems of global climate warming from greenhouse gases.

The Canadian geologist N. W. Rutter points out that glaciers affect the land in many ways. Accumulated glaciers, weighing millions of tons at the height of the Pleistocene Ice Age, bent the land downward under their staggering weight. When the ice sheets rapidly wasted away, seawater flooded in and, over time, covered these areas with fine marine sediments. Now, long after the heavy ice receded, the land is slowly rebounding, lifting the sediment above sea level. The Hudson Bay lowlands is one such area, rebounding approximately 1.1 yard (1 m) per century. The advancing western coastline of Hudson Bay brings new areas above sea level with each passing year. These glacial rebounding areas are readily observable in Wapusk National Park, located near Churchill, Manitoba, and Polar Bear Provincial Park, in northern Ontario.

Some land areas known to be heavily glaciated bear little sign of the effects of glaciation, while others are greatly affected by the buildup and wasting away of glaciers and ice sheets. It is now known that if an ice sheet is thin or develops in a cold, continental climate, the glaciers and ice sheets of the area are cold based (frozen to the ground) and have little or no glacial erosional effects on the surface of the land. On the other hand, an ice sheet that is thousands of yards thick or develops in a warm, marine climate often is warm based (experiences some melting underneath it). A warm-based glacier moves faster at its edges and slides (sometimes surges) across the ground. Warm-based glaciers produce features on the surface of the land through two glaciation processes: glacial erosion (wearing away) and glacial deposition (filling in).

Glacial Erosion

Erosion by glaciers normally takes place by abrasion and quarrying (plucking). Abrasion occurs when fine particles and rock fragments held in the ice near the base of the glacier move across underlying material, commonly bedrock. This process can striate and polish fragments in the ice and on the underlying bedrock. These glacial striations and marks are easily observed on many parts of the Canadian Shield. The scratches and striations left in the bedrock indicate the direction the glacier traveled. In addition, abrasion may form elongated flutings (gutterlike channels) in the bedrock. Quarrying, the plucking of blocks of bedrock by overriding ice, usually occurs where the bedrock is easily fractured, such as where joints in the bedrock are present. *Roches moutonnées* are the results of both abrasion and quarrying. These rock sheep may look like islands in large lakes on the Precambrian shield, large knobs with a streamlined shape—that is, gradually sloping on the up-glacier side but with a steep face that has been plucked on the down-glacier side of the knob, or island. In a large shield lake, at a distance they appear like sheep in an aquamarine pasture.

In the mountains, erosion action by glaciers often forms U-shaped valleys, sometimes resulting in huge glacial troughs. Glaciers pick up boulders—from fist size to house size—and sometimes carry them hundreds to thousands of miles before dumping them on the ground. The deposited boulders have no geologic relationship to the underlying rock formations. Boulders called erratics, found far from their source rock, are scattered across the glacial landscapes of Canada.

Glacial Deposition

Whatever is picked up by a glacier must eventually be put down—either in the same area or in a different area (glacial erratics). The unsorted veneer of rock and soil fragments left behind once the glacier has melted is called glacial till, a nonstratified mixture of soil containing all sizes of debris, from large stones to clay particles. Drumlins are formed when huge mounds of glacial till are deposited on the ground and shaped into streamlined hills by the glaciers. They often occur in sets, called drumlin fields.

As glaciers move across the land, they may bulldoze large masses of glacial till in front of them or along their sides. These ridges are called moraines. Lateral moraines form along the sides of mountain glaciers as the glaciers bulldoze down the valley; medial moraines (the dark lines running down the center of large mountain glaciers) form from the lateral moraines of smaller, higher glaciers that merge into larger, valley-bottom glaciers; terminal moraines are ridges pushed up in front of a glacier and usually mark the glacier's farthest advance.

Lateral and medial moraines and U-shaped side valleys are components of the landscape of the Kaskawulsh Glacier, Kluane National Park and Reserve, Yukon Territory.

An esker marks the path of a stream that once surged through a tunnel in or under a glacier. Once the glacier has melted, the sands and gravels deposited in this tunnel form a long, narrow ridge that snakes across the ground. When sand and gravel are washed into a hole in the glacier, these rock fragments may be deposited as a cone-shaped hill, called a kame. Crevasse fillings are short ridges formed on the land as a result of a crevasse in a former glacier having been filled with sand and gravel. Kettles are round depressions—the reverse of kames and crevasse fillings—and form when a large block of ice breaks off from the main body of ice and is buried in a gravel deposit. Kettle lakes and ponds are common features of the Canadian taiga.

Meltwater channels carry the water from melting glaciers across the surface of the land. After they are dry, these channels become valleys of various sizes. They can be tiny valleys nestled within a boreal woodlands. On the other hand, they may be the immense valleys of the western plains, valleys that are more than a mile wide and several hundred feet deep and often fitted with a narrow meandering stream. Meltwater also formed lakes in front of the ice sheets, some of them several times larger than Lake Superior is today. Meltwater channels brought into these glacial lakes tons of sediments—gravel, sand, silt, and clay—often sorted by size. These became the various soils of these huge glacial lakes and form the rich agricultural soils of the Canadian prairies. Thus the grain farmers of the Prairie Provinces are actually farming a legacy left behind by the Pleistocene ice sheets.

These are just some of the ways that the glaciers have sculptured the surface of Canada. As N. W. Rutter correctly points out, Canada was 97 percent covered by

Thousands of small kettle lakes like this one are found in the boreal forest region of the Interior Plains.

the Pleistocene ice sheets, and as a result this vast land contains more glaciated terrain than any other country in the world. In its past Canada truly was a land of ice and snow.

Appreciating Shield Country

If the boreal forest can be viewed as Mother Earth's evergreen mantle, then the Canadian Shield should be seen as her royal emblem. With its aquamarine jewels held in a granite and forest-green setting, the Canadian Shield is a fitting royal brooch. Its glistening lakes and countless islands offer as breathtaking a landscape as any found on Earth.

In what other region of the world could I watch my young daughter run her fingers across the scars on bedrock and talk to her about the mile-high glaciers that left them there? Where else could a seven-year-old line up her school of "rock turtles" and teach an imaginary class of pupils who are all over 2 billion years old? Where else can a family, canoeing along the bare rock faces, search for stromatolites that are some of the oldest signs of life found on our planet?

The Canadian Shield is a bizarre land yet a basic and quintessential landscape. To understand the Canadian Shield is to understand the primordial processes that formed the North American continent. This is a landscape with a natural history extraordinary enough to challenge anyone's imagination. And yet Canada has very few parks protecting areas of the shield. To most Canadians, it is just an endless tract of rock and stunted trees and an economic wasteland. Perhaps it is time for a different perception.

To my mind the taiga and the Precambrian shield represent a special kind of ecological association, whether we are focusing on the gigantic muskegs of the Canadian Shield, or the flower-laden evergreen forest of the Baltic Shield, or the light and dark taiga forests found on the Angara Shield of eastern Siberia.

The combination of granite rock, spruce, and tamarack—born of bedrock and a frigid boreal climate—is a study in northern geometry that has challenged generations of artists and writers. Images found in the writings of Chekhov, the films of Ingmar Bergman, and the canvasses of the Canadian artist Tom Thomson are derived from this land.

I remember how struck I was by the harsh beauty of the Precambrian shield the first time I ventured onto it. I had lived and worked in the taiga but always on the Interior Plains. On the shield, I noticed a deeper sense of adaptation, of rock and plant and animal fitting together. There, many boreal processes make better sense. Trembling aspen, for instance, which grow on the warmer sites of the Canadian Shield, create their own microenvironments as they patiently send out suckers across bare Precambrian rock. Such integration and balance are the essence of shield country. It often takes the eye of the artist to capture these attributes. Tom Thomson pursued this essence throughout his short, productive life. His canvasses strove to express shield country, the northern angularity of a taiga landscape—a landscape that, if we listen, speaks deeply to the consciousness of all northern people.

General References

Bastedo, J. 1994. *Shield Country: Life and Times of the Oldest Piece of the Planet*. Calgary: Arctic Institute of North America.

Gould, S. J. 1989. *Wonderful Life: The Burgess Shale and the Nature of History*. New York: W. W. Norton.

Kupsch. W. O. 1974. The Churchill-Reindeer Rivers Area: The Evolution of Its Landscape. *The Musk-ox* 10:10–29.

Marsh, J. H., ed. 1990. *The Junior Encyclopedia of Canada*. Edmonton: Hurtig Publishers.

———, ed. 1994. *The Canadian Encyclopedia*. Edmonton: Hurtig Publishers.

Moon, B. 1970. *The Canadian Shield*. Toronto: Natural Science of Canada.

Ojakangas, R. W., and D. S. Darby. 1976. *The Earth: Past and Present*. New York: McGraw-Hill.

Parker, S. P., ed. 1988. *McGraw-Hill Encyclopedia of the Geological Sciences*. 2d ed. New York: McGraw-Hill.

Skinner, B. J., and S. C. Porter 1987. *Physical Geology*. New York: John Wiley and Sons.

Stokes, W. L., S. Judson, and D. Picard. 1978. *Introduction to Geology: Physical and Historical*. Englewood Cliffs, N.J.: Prentice-Hall.

Storer, J. 1989. *Geological History of Saskatchewan*. Regina: Saskatchewan Museum of Natural History.

4

Seasons to Burn

Over much of the North American taiga, winter brings thick snow, short days, clear cobalt-blue skies, and a low, amber-colored Sun glancing off an impeccable whiteness. Summer brings warm, long days and an absence of ice (above ground, at least), luxuriant green growth, swarms of insects, and a forest floor and shrubs sprinkled with flowers, sometimes followed by an abundance of berries. In the taiga year, summer is a brief cornucopia of reproduction and growth. Spring and fall are major transitions, times of change for both earth and sky.

Many indigenous cultures of the taiga and the tundra recognize these familiar seasons. However, there is no reason that the seasons should be limited to four. The Eastern and Woods Cree, who occupy the boreal forest from northern Quebec across into northern Alberta, divide the year into six seasons: *sikwan* (spring), *mithoskamin* (breakup), *nipin* (summer), *takwakin* (fall), *mikiskaw* (freeze up), and *pipon* (winter). Four of these seasons are familiar to us, two of them are unfamiliar—and interesting to explore.

Mikiskaw occurs during late fall and early winter, following the brilliant autumn colors and the last of the berries. It is a time of major transition from the spectacular yellows and golds of takwakin into the radiant whiteness of pipon. During mikiskaw the land is forced into a major metamorphosis, a struggle between the cold air and the warm soil in the increasingly long nights of October. It is a season of leaden skies and strong winds, a time when the rivers and lakes, with their cresting waves and near-freezing water, become exceedingly dangerous to travel by boat or canoe. It is a period of waiting: The leaves have been shed, and even the antique-gold slivers of tamarack needles have tarnished and fallen. It is a time for letting go, for surrendering to winter.

Mithoskamin is another major earth-sky transition. The metamorphosis begins when summer ground appears under encrusted snow, until the lakes are unsealed, transformed into their summer selves. In the Woods Cree country of the northern Prairie Provinces, mithoskamin extends from late March until late May or early June. It is a transition orchestrated by the Sun, which shines for twelve, then four-

Tamarack and other members of the larch family are deciduous conifers. Their needles turn gold and shed during autumn.

teen, and then sixteen hours each day. Solar radiation finally overpowers the delicate structure of snow. This white blanket, which has covered and protected the ground during the past six or seven months, begins its final disintegration. Solar warmth melts the surface of the snow, and water percolates down, dissolving the delicate, crystalline structure of the snow. Moisture collects on the surface of the ground, and the snow cover begins to collapse from the ground upward. When the structure of the snow deteriorates like this, it is a time of immobility for all large mammals, including humans, except at night or in the early morning, when the cold night air has temporarily stiffened the nivean structure. But soon even this strategy fails, and moose, wolf, and trapper alike face two weeks or so of restricted movement.

The Stresses of Spring

In the land of the Eastern and Woods Cree, by early May the snow is often nearly gone, but the ice on the lakes remains thick. It has been building throughout the winter, until now on the large, windswept lakes, it usually is still three feet (a meter) or more thick. Also at this time the ground may be frozen so that the roots of plants are embedded in frozen soil. The deciduous trees have not put forth their leaves, the strategy of the northern races of aspen, balsam poplar, and white birch, which have to cope with the drought of spring.

(Note that my use of anthropomorphisms is a way of summarizing an evolutionary process; the foregoing could also be stated scientifically: Over time, certain deciduous tree species randomly exhibited mutant phenotypes of delayed leaf appearance. Given the taiga's spring drought, these phenotypes were selected for, and individuals showing them passed more of their genes on to the next generation than trees that leafed earlier. The survival value of this phenotypic trait was great enough that it eventually became established as a common inherited feature of the northern race of this tree species.)

Mithoskamin is often marked by abundant sunlight, few clouds, low humidity, and warm, balmy winds. Given these dry conditions, the season is a time of great stress for trees, particularly conifers, which not only lose the small amount of moisture they contain but also gain very little moisture from the frozen soil. Furthermore, the lakes—the local rainmakers—are still sealed with ice. One can see this water stress best in the pines, whose needles often become a sickly yellow-green. The moisture problem provides a plausible explanation for the deciduous needles of the tamarack and other members of the larch family. It may be an evolutionary tactic: Larches shed their needles in the fall to avoid the water stress of mithoskamin. Since tamarack often grows in bogs with a thick insulating layer of sphagnum moss, and other larches often grow high on mountain slopes, they occupy some of the last habitats to thaw after winter. The loss of their needles may be an adaptation to habitats subjected to severe spring drought.

Because of its drought conditions, mithoskamin can also be a major fire season in the boreal forest. The conifers are dry; there is little or no fresh green growth on the forest floor, the lichen and the dried stems of cattails, bulrushes, grasses, and sedges are flammable. Mithoskamin seems to progress through an unrelenting series of fateful events that reminds me of Shakespeare's *Romeo and Juliet*. Both begin with a festive atmosphere and move inexorably toward destruction. By mid-February the days have become luxuriously long, and by March much of the taiga is gaining six minutes of sunlight each day (even more farther north). Mid-March, with its long cloudless days and abundant sunlight, is a time of solar revelry. However, there will probably be no rain for the next two months—until the large lakes are ice free. While the brilliant, Sun-saturated days continue one after the other, the trees get drier and drier. Then one day in May the sunlight has a slight silvery cast. Soon other forebodings occur—shadows have a harsh edge, the moon throws a dingy light. Somewhere, perhaps hundreds of miles away, a fire has begun.

Smoke from a forest fire often rises high into the atmosphere on its own heat, and carried by winds it can arc several hundred miles before returning to ground level. As a result, when smoke is detected in the air, it is difficult to know if the fire is thirty or three hundred miles away; it is difficult to know what the threat is to your local forests or your home community. All you are sure about is that somewhere the destruction has begun. Days, even weeks, can be filled with smoke. You can taste it, you can feel it in the back of your throat, your lungs. People modify their behavior, their desire for exercise evaporating. You wait and watch, wondering how close the fire will come. You begin to hope, and more than hope, for the life-giving rains that will cleanse the pallid air.

The smoke of mithoskamin affects many things in the forest. One May evening as I watched a soiled moon rise, a pack of wolves called from across a small meandering river. Over the years I have heard wolves howl many times, but I had never heard them call as they did on that smoke-laden night. Wolf biologists believe that howling advertises that a particular territory is occupied, reinforces the bonds among pack members, and communicates with offspring back at the rendezvous site. But is more going on? Is there an emotional or aesthetic element, which science misses? On that smoky evening I thought I could hear the soreness in their eyes and lungs. I thought I could hear apprehension.

The mithoskamin fire season is a fact of life for much of the taiga across its intercontinental span. Only coastal areas, which are influenced by the never-freezing oceans, seem to escape this season of burns. Mithoskamin is not the only time of a high fire hazard. Whenever several weeks of little or no rain coincide with warm, dry winds and dry lightning storms, a high danger of fire develops. Furthermore, the weather systems that cause the dry climatic conditions of August and that cure grain crops across the prairies may extend unduly northward. If so, the summer taiga, despite its luxuriant green foliage, may experience a fire season. None of these fire seasons occurs like clockwork. Some years a fire season develops and some years it doesn't. But over time, over the lifespan of trees, fires are a fact of life in the taiga. Forest fires determine important patterns in much of the North American taiga: Although certain fire-prone areas (such as sandy ridges and rocky outcroppings) burn frequently and other sites (such as islands or floodplains) are virtually immune to fire, on average (with variation among sites), most of the taiga probably burns every 100 to 200 years. This is the documented fire cycle for much of the North American taiga, and it is this fire cycle that has long determined many of the plant dynamics of the boreal forest.

To understand fire's role in shaping boreal ecology, it is important to realize that the circumstances that create much of the Canadian taiga are different from those that create most forests. The rain forest of the northern Pacific coast results from the abundant moisture that sweeps in off the ocean. Abundant moisture also supports the deciduous forest that cloaks the eastern half of the contiguous United States. This latter area receives an average of thirty to forty inches (75–100 cm) of precipitation annually. But the North American taiga over a good portion of its range receives as little as sixteen inches (40 cm) of precipitation per year—not much when one considers that even deserts usually get ten inches (25 cm). The only reason a forest exists in northern Canada is that this stingy amount of precipitation is combined with an equally parsimonious amount of evaporation. It is the positive balance between "water in" and "water out" that allows the trees of the taiga to grow. Because minimal water evaporates from land and water surfaces and because, up to a point, plants control the amount of water they transpire, the boreal forest can exist on its frugal water budget. Scientists call such water loss *evapotranspiration*. The fact that the precipitation, though restricted, still exceeds evapotranspiration is a key ecological process that allows the forest vegetation of the taiga to prosper.

A number of factors reduce evapotranspiration in the boreal forest: the short growing season, the cool nights, the frequent cloud cover, and the Sun's low angle of incidence. In essence, the low level of evapotranspiration puts this land into a positive water balance and permits the vegetation of the taiga to grow. However, it is a fragile balance, which can be easily disrupted. In years when there is thick ice on the lakes and a large high-pressure system lodged overhead, a mithoskamin fire season develops, and large tracts of the boreal forest burn. The pattern has been repeated for thousands of years, but it is a pattern that can be intensified by the drying and warming effects of global climate change and other human-related activities.

Fire as the Driving Force

Plant communities develop through a process called *succession,* which involves changes over time in species composition and in the structure of the plant community. Most plant communities are subject to natural disturbances—fire, insects, disease, avalanches, floods, windstorms. These disturbances kill many plants of the forest and reset successional processes by which the forest renews itself. In the taiga, forest fires are the dominant short-term destructive agent. Historic records show that some areas of the boreal forest burn as often as every 50 years. Few boreal forest stands reach more than 200 years of age. Because of these frequent fires, the boreal forest often shows large areas of even-aged stands, composed mainly of early successional trees and shrubs established after the most recent fire.

Fires are very complex in the way they burn and in their ecological effects. These fire regimes and the equally complex postfire responses of boreal plants result in a mosaic of habitats, vegetation types, and successional stages occurring on the landscape. For example, several boreal tree species are adapted to reproducing shortly after an intense forest fire. Some conifers (jack pine, lodgepole pine, and black spruce) keep their cones on their branches for many years. These cones (serotinous cones) contain viable seeds but are sealed until opened by the heat of a fire. When the seeds are released, they germinate readily on the mineral soil exposed by the fire. (We look at this adaptive process in detail in the next chapter.) Aspen has a different reproductive strategy: even though its above-ground parts may be killed by an intense forest fire, most of its roots survive and produce suckers along the length of its roots. Paper birch usually produces sprouts growing up around the root collar, resulting in the familiar clusters of multistem birch forming around a fire-killed trunk. Understory plants can survive as seeds buried in the soil for decades, only to germinate when the canopy and leaf litter are removed by fire and Sun warms the soil.

Fire Profiles

To understand these plant dynamics, we begin by investigating some of the complexities of forest fires. Many people have the impression that a forest fire is a destructive event that totally consumes the forest in its path. Not all forest fires are like that. They are very complex in their behavior and even more complex in their ecological effects. Forest fires can be grouped into three broad types: surface fires, crown fires, and ground fires.

Surface Fires

Surface fires occur in the boreal forest only under certain conditions: moderate moisture conditions before the fire and moderate wind conditions during the fire. Under these conditions, the fire burns only across the surface of the ground; it does not crown out. A surface fire consumes shrubs, herbs, dried grasses, and other ground vegetation in addition to fallen leaves, needles, twigs, and branches. If a surface fire develops sufficient heat or burns in one area long enough, it may heat girdle and kill thin-barked trees, but if it burns only across the surface of the ground, it generally won't lead to the wholesale destruction of a mature forest. Furthermore, the high moisture content of the green summer foliage of herbs and shrubs eliminates a critical energy buildup, thus preventing a surface fire from escalating into a crown fire.

A surface fire often thins out the forest because it kills some saplings, shrubs, herbs, and even weaker trees. It may also reduce combustible fuels within a stand to a safe level, thus helping to ensure against future fires. A forest stand that periodically experiences a surface fire is markedly less flammable than a forest in which light fuels (twigs, lichens, dried grasses) have accumulated for years. Thus periodic surface fires are the best insurance against a wholesale forest holocaust. After a surface fire, the root systems of shrubs and trees are usually still alive. Shrubs, deciduous trees, and on rare occasions even conifers are able to sprout, so that three or four years after a surface fire there are few visible signs left of the fire.

Crown Fires

A crown fire occurs when a surface fire builds up enough heat to ignite the overlying canopy. White and black spruce, and to a lesser extent balsam fir, because of their high-density branches and their characteristic of retaining dry, dead twigs, often act as ladder trees, allowing the fire to climb into the canopy. (In Chapter 6 we explore why these trees develop such fire-prone characteristics as they mature.)

Given dry conditions and strong winds, a crown fire can build quickly into a major conflagration. A description of such a fire that occurred near Lesser Slave Lake in northern Alberta during the spring of 1969 shows that, once a fire crowns, strong winds fan it and sweep it along. This particular fire moved at four miles per hour (7 km/hr) over a ten- to fifteen-mile (15–25 km) front, killing most of the trees before it. During one ten-hour period it burned 150,000 acres (61,000 ha) of forest—an area the size of Crater Lake National Park in Oregon. It is calculated that the fire released the energy equivalent of a megaton bomb every ninety minutes. At or behind the fire front, the wind force (called the *firestorm*, largely generated by the fire itself) was strong enough to uproot trees and snap off trunks ten or fifteen feet (3 to 5 m) above ground level. Elsewhere, the firestorm created giant whirlwinds, which uprooted and scattered thousands of trees in bizarre circular patterns several hundred feet in diameter. Yet because this was a spring crown fire and the ground was still frozen, the roots and seeds of many plants survived. In fact, one month after this severe fire, the burned area was green with vegetation. Had the same fire occurred during summer or autumn, the entire organic layer of the soil, together with the root systems of trees and shrubs, might have been consumed.

Ground Fires

A ground fire behaves in a completely different way from a surface fire or a crown fire. A ground fire, more precisely labeled an *organic soil fire*, is a glowing combustion that becomes established in organic soils, such as the deep peat deposits found in northern muskegs. Ground fires may smolder for days, weeks, even seasons before they break out on the surface to be fanned by winds into full-scale crown fires.

These fires have some unusual aspects. For example, a ground fire in a peat bog tends to burrow into the peat deposit, burning most intensely and spreading faster at the lower levels. As smoke and gases rise through the dry peat, they partially suppress fire in the upper peat levels. This burrowing characteristic makes detection and mapping of a ground fire difficult. Ground fires can also be exceedingly difficult to extinguish. Contributing to this difficulty is that as sphagnum moss smolders it gives off paraffin gases. As these warm gases rise, they cool and condense as a coating of water-repellent paraffin (a waxy crystalline hydrocarbon) on the upper layers of peat, making it resistant to penetration of water. Saturating these upper layers of peat with water thus becomes one of the most difficult aspects of fighting a ground fire. Russian scientists are experimenting with water-injection devices and wetting agents, such as sulphanole, that cut through paraffin. This research should aid in fighting ground fires more effectively throughout the taiga.

Ground fires also can cause multiple burns in a forest, meaning that the same area is burned repeatedly within a limited period. The sequence is something like this: An initial fire kills many of the trees and shrubs in an area without actually consuming them (green wood does not burn well). At the same time, it ignites a ground fire in one of the muskegs. Over the summer, the fire-killed trees and shrubs dry out, and toward the end of summer the ground fire surfaces and ignites a second and more destructive fire. In this second fire, much more of the organic soil and plant communities is consumed. This sequence may be repeated a number of times until the area is left with few viable sprouts and a poor supply of dormant seeds. After multiple burns, the area will have an almost sterilized look, and the vegetation will be slow to recover after the fire finally goes out.

These fires may also be responsible for one of the oddest landforms in the taiga. These are boulder fields, or cobble streets—large tracts of scrubbed boulders laid down next to or on top of each other in the midst of undisturbed forest. These cobble streets may be several hundred yards in length. They may contain side streets extending off the main area. They are a strange sight to come upon unexpectedly in the midst of a densely vegetated forest, and I am indebted to Walter Lyford, of the Harvard University Forest, who first showed me a boulder field years ago. Such landforms are not common. I have stumbled upon only a couple of them in my years of bushwhacking in the taiga. One that can be hiked to fairly easily is in the southwestern corner of Prince Albert National Park, in Saskatchewan.

What causes boulder fields to form in the middle of a forest is unclear, and theories are still debated. One hypothesis attributes them to ground fires that have burned organic terrain rather completely. This hypothesis suggests that many years before the ground fire began, frost action caused boulders and large stones to work their way up from the underlying mineral soil to the top of the mineral soil or even into the peat. This frost-heaving action may have been going on for hundreds of years. Then one or several ground fires became established and consumed the peat almost entirely. The boulders, scoured clean by the heat of the fire, were then let down onto the mineral soil, forming the cobble street where the organic terrain used to be. Due to the action of wind, rain, snow, and perhaps subsequent fires, many years may pass before enough organic soil develops on these boulders to support plant growth.

Once established, ground fires can demand a Herculean effort to extinguish. Anyone hiking, canoeing, hunting, or traveling in the taiga should be extremely careful never to build a campfire on peat or on any deep organic soil because of the danger of starting one of these insidious conflagrations. The traditional way of fighting ground fires is to dig a trench down to the mineral soil, making sure that the trenched area encloses all the points of glowing combustion. The entrenched area is then deluged with great quantities of water until the fire is out.

Phillip Bird, a Cree from northern Saskatchewan and an experienced firefighter, told me about a smoldering ground fire near Montreal Lake, Saskatchewan, a mile or so from his home community. His crew of firefighters tried to dig a trench to contain the ground fire, but they kept encountering branches of the smoldering fire that had tunneled into new areas of the muskeg. Time and time again, the fire crew had to enlarge their system of trenches. They dug each trench down to the mineral soil, often twelve or more feet (4 m) through the peat. The trench frequently filled with water and had to be pumped out so that the team could continue to dig deeper into the water-saturated morass. They continued in this quagmire work for six weeks before the trenches entirely enclosed the fire. That was the only way they knew to protect their homes and school from the spreading ground fire.

Summary

Research shows that every century or two most areas of the North American taiga are consumed by forest fires and that this pattern has been going on for thousands of years. To fully understand and appreciate the role of forest fires and their effects on the plants and animals of the boreal forest, we must explore these dynamics from an evolutionary perspective. The next chapter poses some questions about the evolution of taiga species, and then it looks closely at the adaptive responses of two species to the destructive forces of forest fires.

General References

Heinselman, M. L. 1973. Fire in the Virgin Forests of the Boundary Waters Canoe Area, Minnesota. *Quaternary Research* 3:329–82.

———. 1973. The Ecological Role of Fire in Natural Conifer Forests of Western and Northern North America. *Quaternary Research* 3:317–28.

Kelsall, J. P., E. S. Telfer, and T. D. Wright. 1977. *The Effects of Fire on the Ecology of the Boreal Forest, with Particular Reference to the Canadian North: A Review and Selected Bibliography.* Occasional Paper 32. Ottawa: Canadian Wildlife Service.

Johnson, E. A. 1992. *Fire and Vegetation Dynamics: Studies from the North American Boreal Forest.* Cambridge: Cambridge University Press.

Rowe, J. S., and G. W. Scotter. 1973. Fire in the Boreal Forest. *Quaternary Research* 3:444–64.

Shugart, H. H., R. Leemans, and G. B. Bonan, eds. 1992. *A Systems Analysis of the Global Boreal Forest.* Cambridge: Cambridge University Press.

Theberge, J. B. 1998. *Wolf Country.* Toronto: McClelland and Stewart.

Viereck, L. A. 1973. Wildfire in the Taiga of Alaska. *Quaternary Research* 3:465–95.

Wein, R. W., R. R. Riewe, and I. R. Methven. 1983. *Resources and Dynamics of the Boreal Zone.* Ottawa: Association of Canadian Universities for Northern Studies.

Wein, R. W., and D. A. MacLean, eds. 1983. *The Role of Fire in Northern Circumpolar Ecosystems.* New York: John Wiley and Sons.

5
The Benefits of Being Burned

One of the attributes of living organisms is the ability of one species to slowly, through evolution, turn an adverse interaction with another species into a beneficial relationship. Consider predation. A predacious lifestyle appears to be basically selfish: A predator benefits by devouring a prey. The predator gets a full stomach; the prey is destroyed. It seems clearly exploitative, and yet, through evolution, so many secondary benefits accrue to the prey population that it is hard for the biologist to judge whether the prey population or the predator population gains more from this apparently selfish act. In some instances, both benefit—predation becomes an effective population-regulating mechanism not only for the prey but also for the predator.

There are other benefits from predation that accrue to the prey. Predators may remove diseased animals before an infection or parasite has a chance to spread through the prey population. Predators can also keep a prey population young and vigorous by removing the injured, old, and feeble individuals. Certainly, over eons, pressure from predators selecting the least fit prey has slowly developed the alertness, agility, and speed of various prey species. Hunting by lions and cheetahs is largely responsible for the grace and speed of Thomson's gazelle. Hunting by wolves, coyotes, and humans has made white-tailed deer wary and fleet footed.

The evolution of adverse interrelationships into mutually beneficial ones can be seen in other biological relationships as well. Consider bees and flowers. Originally, bees were probably flower pilferers, scavenging nutrients they happened to find in blossoms. But over time flowers took advantage of this thievery, adapting their reproductive systems to make use of it. No longer would flowers cast their pollen on the fickle wind; instead bees, enticed by nectar, would carry the pollen from one flower to the next, ensuring cross-fertilization and promoting genetic variation in the plant species. In return, the bee would receive an abundant supply of nectar. This relationship has evolved to such an extent that the full design of many flowers becomes visible only when we view these flowers through the bees' eyes—that is, through filters that transmit ultraviolet light. Furthermore, the stamens of some flowers have evolved elaborate triggering devices to ensure that foraging bees are dusted with pollen, and certain species of bees have evolved pollen-collecting combs to transport pollen more effectively. This coevolution of flowers and bees has resulted in obligate dependency for both.

Is it possible that this same dynamic—the transformation of seemingly adverse interactions into mutually beneficial ones—is at work not only between species but also between species and their environment? In the case of the taiga, is this dynamic at work in the interactions between the plants and animals of the boreal forest and its main destructive force—forest fires?

Most of the taiga is exposed to periodic burns, sometimes of great intensity. Consequently, over eons, many boreal species of plants and animals have "experimented" with making this holocaust beneficial. The results of these evolutionary experiments are diverse and intriguing. In this chapter, I examine the adaptations that one animal species, the moose, and one plant species, the jack pine, have evolved in response to forest fires—how these two species have been sculpted by the evolutionary selection pressures of forest fires.

Moose

The moose is the largest member of the deer family, and its distribution in Canada parallels that of the boreal forest. In fact the moose and its habitat are so closely interrelated that a nickname for the boreal forest is the moose-spruce forest—a most appropriate epithet.

A moose is about the size of a horse; males weigh between 900 and 1,200 pounds (400 to 550 kg); females are somewhat smaller, between 600 and 800 pounds (270 to 360 kg). The average bull moose stands about six feet (2 m) high at the shoulders and more than eight feet (2.5 m) tall to the top of his antlers. At close range, a bull moose is truly an imposing animal. I remember one evening when I was quietly hiking along a narrow trail, hurrying to get back to camp before dark, and in the twilight startled a bull moose lying down only a few yards off the trail. The mammoth animal leapt to its feet, lowered its antlers, and charged toward me. It then pivoted and went crashing through the dense bush, while I stood quivering, listening to shrubs and saplings snap as this huge, dark creature raced away. The encounter left me deeply impressed with the size and power of *Alces alces*.

Moose are solitary animals except when they are raising young or courting. The calves are born during early June and stay close to their mothers for their entire first year. A strong bond develops between cow and calf, although not immediately. Shortly after birth, a calf instinctively follows any nearby moose, usually its mother. For the first few weeks of the calf's life, the mother drives off any other moose in the nearby area. Unlike deer fawns, moose calves do not spend a large portion of their early life lying down, hiding in vegetation. It is as if natural selection has decided that, when you have a mother this big, you don't need to hide. Instead, the moose calf follows and stays close to the cow, which threatens humans and other animals by holding her head low, giving guttural vocalizations, and bristling the hair on her neck and shoulders. She may strike out with her front hooves or kick like a mule with her hind feet, skills evolved by moose to a fine degree. Wolves are occasionally found with broken jaws or ribs, which are attributed by biologists to encounters with moose.

If female moose winter on high-quality food, such as recently burnt-over areas, a significant number will produce twins. If their winter diet is less nutritious, they will likely give birth to a single calf. The calves are small at birth, from about thirteen pounds (6 kg) each for twins up to perhaps thirty pounds (14 kg) for a single birth. However, they have the highest growth rate of any North American ungulate and may put on more than four pounds (2 kg) per day. Their mothers can consume up to thirty pounds (14 kg) of fresh spring foliage per day, which when transformed into rich moose milk supports the rapid growth of their offspring. The calf is weaned at about three months of age but stays close to its mother through the summer, the rutting season of late September, and over the winter. The bond between the two is so strong that during the rut the cow will normally reject a bull that shows any aggression toward her calf.

Moose Calling

Although moose are usually silent, vocalizations play an important role during courting. As she enters her receptive period, the cow makes a low, resonant sound, which can be heard by male moose more than a mile away. Bulls are understandably attentive to the bawling of the females as well as to other sounds, such as the noise of antlers rubbing against the branches of trees or shrubs, which may mean they have a rival in the area. The receptivity of bulls to these sounds is often taken advantage of by moose-hunting humans, who have raised moose calling to an art, employing birch bark horns and other devices. The sound can also be approximated with a quart-sized (liter) tin can. Cut one end out of the can and punch a hole in the other end. Tie a knot in a stout cotton shoelace and string it through the hole, letting it dangle from the bottom of the can. Now wet the lace thoroughly, pinch it between two fingers, and run your fingers down the lace. The bawling sound that results is fairly close to the call of an amorous female moose.

The most unusual moose-calling method I have encountered was developed by the Gwitchin Indians of the Yukon and eastern Alaska and is used throughout this region. They have devised a moose-bone technique, which the anthropologist Richard Nelson describes. The Gwitchin use a carefully dried shoulder blade from a cow moose (the bull's scapula is too heavy). To get the proper sound out of this instrument, the bony ridge running down the middle of the shoulder blade is cut down low, the cartilage edge of the bone is left in place to dry, and the joint at the other end is carved into a handle. While hunting, the Gwitchin drag the moose bone over shrubs or small trees with a brisk but gentle action, not unlike that used with a paintbrush. The sound mimics that of moose antlers rubbing against branches, and the Gwitchin prefer this method of moose calling because they say it is useful for more than just the couple of weeks when the females are bawling. After the rut, bulls tend to congregate into small groups, called *pods*, and playfully spar with each other in antler-pushing contests. The Gwitchin say that the moose-bone technique can attract bulls throughout the podding season of early winter.

Moose Habitat

To understand how fire in the boreal forest affects moose, it is necessary to understand how moose populations are organized. Because of their diet, moose do not do well in a mature conifer forest. They depend upon aquatic vegetation and deciduous

The Gwitchin moose-calling device is fashioned from the shoulder blade of a female moose.

foliage as mainstays of their summer diet and upon deciduous browse—the twigs and buds of willow, birch, aspen, rose, and other leafy plants, with some balsam fir tossed in for good measure—for their winter diet. Balsam fir is the only conifer species used to any significant extent by moose. Thus it is common during winter to see moose associated with dense stands of deciduous shrubs or saplings or dense stands of young balsam fir. Furthermore, wintering moose are often associated with the dense deciduous growth that sometimes comes up after a forest fire. Recall that when a section of taiga is burned without damage to the roots, the stand often regenerates into a dense growth of suckering aspen, birch, willows, and other sprouting shrubs, providing moose with an excellent and abundant food supply. Moose will invade a burned area a year or two after the fire. Since deciduous browse is abundant in such areas for a number of years, the cows frequently twin, and the moose population expands rapidly, remaining dense for ten or more years.

Over several decades, however, the forest matures, and conditions change. Young aspen, birch, and willow—Sun-loving species that need full sunlight to germinate and grow vigorously—do not do well in the shady conditions under an established canopy. In fact, aspen and birch seldom germinate at all under the shade of their parent trees. Instead conifers, mainly white spruce and balsam fir, establish themselves under the canopy, gradually replacing the deciduous tress and shrubs. Also, as aspen, birch, and shrubs grow taller, much of the new growth—the most nutritious browse—grows out of the moose's reach. Especially in winter, moose make considerable effort to reach this browse. They straddle shrubs or stretch their necks so that they can reach as high as nine feet (3 m) above the ground in order to grab the tops of saplings. They also push against the saplings to break them so they can feed on the twigs and buds, which are the most nutritious. Broken shrubs and saplings, particularly young birch with broken tops, are a common field sign left by moose during late winter. However, in about fifteen or twenty years, as the deciduous trees and shrubs grow out of reach and young conifers take over, the moose population begins to decline. Until there is another fire, the stand will eventually support only a few of these animals.

In addition to postfire areas, some moose inhabit "permanent" habitat—areas that on a relatively permanent basis maintain a rich growth of deciduous shrubs and saplings. These habitats might lie along watercourses or in deltas, along small meandering creeks or on mountain avalanche slopes—any area, in fact, where soil moisture conditions, spring floods, avalanches, or other events ensure the vigorous growth of deciduous shrubs or saplings that will not grow out of the reach of the moose.

As the ecologist Val Geist explains, the moose that live in these permanent habitats are the fixed nuclei of the population. There usually aren't a great many such sites in the boreal forest, but those that do exist consistently support a small number of moose. From these sites each spring the yearling moose, both males and females, disperse in search of temporary postfire habitat or, more rarely, vacant permanent habitat. Once driven out by their mothers, yearling moose often travel long distances. They set off in late May or early June (shortly before the new calves are born) and travel until they either perish or find vacant habitat, usually areas that burned less than a decade earlier and that occur here and there in the taiga. Those lucky enough to find a large, freshly burned area become the founders of a subpopulation of moose, which will prosper and inhabit this tract of postfire forest for the next fifteen or twenty years.

In summary, the boreal forest always seems to have a small amount of permanent habitat capable of supporting a stable, if restricted, moose population. But to flourish, the species needs fire—fire to create burnt-over habitat rich in deciduous browse, which will support these giant herbivores for a decade or two. It is this relationship that makes the moose a fire-dependent species.

Jack Pine

The jack pine is a scrappy tree. The species does not have the massive trunks of its majestic cousin the eastern white pine *(Pinus strobus)*. The jack pine was never called the king's pine, nor did it ever supply masts for the tall ships of the Royal Navy, as white pine did. In fact, jack pine is largely considered a weed species, a tree often ignored until its more imposing relatives are overharvested. Only as lumber operations moved west and north in search of new resources did foresters develop an interest in jack pine. Today, this humble pine provides us with construction lumber, utility and telephone poles, fence posts, mine timbers, railroad ties, and Christmas trees. Together with black spruce, it is a mainstay of the pulp and paper industry. Not bad for a weed species. What jack pine lacks in majesty it makes up for in tenacity, growing on sites too dry to support most other tree species.

Because of its dry, sandy sites, the jack pine is predisposed to fire. However, its self-pruning and self-shedding characteristic greatly reduces the risk. By shedding most of its dead branches, it reduces the buildup of combustible fuels in the canopy.

Meandering small creeks create productive permanent habitat for moose.

By self-pruning low on the trunk, it removes ladder branches, which can carry a surface fire up into the crown. The jack pine also sheds dead needles, further reducing crown fire fuel. When jack pine burns, however, it does so with alacrity. The needles have a high resin content, and when water stressed—as they often are during spring and summer droughts—they are highly combustible. Furthermore, the flaky bark, which offers some protection against light surface fires, contributes considerable fuel to a major conflagration.

Jack Pine Fire Strategies

When an intense crown fire burns a site, the leaf litter and even the soil's organic layer are largely consumed, so that in many places mineral soil or bare rock lie exposed. Most of the trees in the stand are killed, and the canopy is largely destroyed. Shade is reduced, and sunlight pours in, further drying out the soil. Since there is no effective canopy, the site is also fully exposed to cool northern nights and the damage caused by unseasonably early or late frosts. These are some of the conditions with which the next generation of trees and other plants must cope.

The kind of postfire forest that then develops depends upon a number of factors, an important one being the state of the soil during the wildfire. If the ground remains frozen or wet during a fire, as often happens during a spring fire, and if there was aspen in the previous forest, then the aspen roots will sucker, sprouting not just around the collar of the original trunk, as paper birch and other hardwood tree species do, but also along the surface roots as they travel out from the trunk. New aspen shoots have been observed by the Alaskan ecologist Leslie Viereck more than ninety yards (83 m) away from the original trunk, an adaptation that makes aspen a successful competitor in recolonizing after a spring fire.

This clone of aspen sprouts—all of which are capable of developing into mature aspen trees if conditions are favorable—are genetically identical, share a common root system, and biologically constitute a single living organism. Aspen clones are easily identified during certain times of the year: They leaf out at the same time in spring and turn golden and drop their leaves simultaneously in autumn. Biologists consider certain of these large aspen clones the largest living organisms on Earth. It is conceivable that aspen clones are also among the oldest living organisms on Earth.

Jack pine seedlings seldom can compete with healthy aspen sprouts, because aspen suckers are supported by a mature root system. But suppose that the fire is more intense. Suppose that the ground is not frozen and that the organic layer of the soil and many of the roots in it are consumed during the fire. After that type of fire, jack pine seeds are often able to germinate and to become the dominant trees of the new forest.

Stan Rowe and George Scotter, the Canadian plant ecologists, rank the trees of the boreal forest according to an index of success. They survey the plant features relevant to establishing after a fire and rate tree species according to these features. In their analysis, of all the conifers of the taiga, jack pine emerged as best adapted for establishing after a fire. Only lodgepole pine, a close relative, can be considered on a par with jack pine. What features does jack pine show in response to fire?

Remembering the conditions that often exist after a severe fire, consider that jack pine seeds germinate most successfully on either bare mineral soil or mineral soil covered by light humus—and that a layer of leaves or a thick layer of organic duff often inhibits their germination. Furthermore, jack pine seedlings grow best under full exposure to the Sun (they are not a shade-tolerant species), and jack pine saplings are capable of surviving drought conditions for a month or longer. In addition, young jack pine trees can be exposed to a sudden drop in temperature (to between 23° and 25°F [−4° and −5°C]) without sustaining much damage. Not only are young jack pines able to cope with drought and frost but also, if given favorable conditions, grow quickly, adding up to fourteen inches (35 cm) to their stems each year.

Another interesting feature of jack pines is that they produce cones at a very early age, sometimes by their fourth or fifth year, often showing quite a number of cones by their seventh or eighth year. Rowe and Scotter point out that these saplings could, in theory, use the energy and nutrients invested in growing cones to produce longer stems to outreach shrubs and other trees with which they must compete. Yet they seem to hedge their bets by producing full-sized cones on relatively young stems. Why? A recently burnt forest often contains dead, drying snags, dry organic soil, partially burnt branches, and other fuels—all of which contribute to the chances of a fire occurring again in that same area. Furthermore, a smoldering ground fire can often surface and ignite when conditions are ideal for burning. Jack pine saplings establish a cone crop early as an evolutionary strategy in case fire strikes again in the same location.

Perhaps jack pine's most impressive fire characteristic is cone serotiny—the ability of trees to retain cones filled with viable seeds for years. In experiments, more than half of the seeds from cones more than twenty years old were able to germinate. Only the heat of a blaze will open jack pine cones, allowing burnt-over areas to be dusted with a layer of jack pine seeds.

Jack pine cones even look unique. As hard as a piece of iron ore, with a surface reminiscent of a hand grenade, they consist of hard segments bonded tightly by a strong resinous glue. Between and behind these hard surface segments lie the compartments containing the seeds. The resinous glue seals off the compartments, protecting the seeds from rain, frost, bacterial and fungal infections, and even the gnawing teeth of most rodents. Red squirrels do persist and occasionally feed on jack pine cones, but given the choice, squirrels favor the soft, fleshy cones of black and white spruces.

Jack pine cones do not explode when exposed to a flame; rather they bloom like a flower filmed in time-lapse photography. The strong glue begins to melt when the temperature exceeds 122°F (50°C). But much more happens in the cone than just the melting of the bonding material. When exposed to heat, the central spine of the cone, to which all the hard surface segments are connected, curls back onto itself, and the openings of the seed compartments blossom into view.

However, the architecture of a jack pine cone has evolved even further than that; the cone also has an ability to protect the seeds from the damaging heat of fire. In the early 1960s, W. R. Beaufant experimentally exposed jack pine cones to extremely intense heat and found that seeds survived when the cones were exposed to 1,650°F (900°C) for as long as thirty seconds. Cones held in a flame at 1,300°F (700°C) for as long as three minutes also showed no loss of seed viability. How is this protection accomplished? Jack pine cones' slightly corky interior provides good insulation, protecting the seeds, at least for a limited time, from damage by fire.

As a strategy against forest fires, jack pine and lodgepole pine produce cones at an early age, often by the time the trees are ten or fifteen years old.

Although cone serotiny is an intriguing example of adaptation to fire, jack pine is not locked into this reproductive strategy. Studies show that about 90 percent of the cones produced by jack pine in the taiga open only in response to the heat of fire. The rest open under the warmth of the Sun. Is this just inefficiency on the part of jack pine? Or does it represent evolutionary wisdom? Many researchers believe that the 10 percent of jack pine cones that open when heated by the Sun represents an effort by jack pine to hedge its bets by using a mix of reproductive strategies. A small percentage of jack pine seeds are released every year, but the majority are banked until a major forest fire calls them forth. When there is a run on the seed bank, the results are impressive. After a typical fire in the taiga, as many as 2 million jack pine seeds per acre (5 million per ha) can be found on the surface of a burnt area.

In the Great Lakes states, the southern part of jack pine range, where the frequency of forest fires is much lower than in the taiga, the majority of jack pine cones open in response to the heat of the Sun. Only a small percentage of jack pine trees produce truly serotinous cones. Thus the reproductive strategy of the jack pine appears to be fine-tuned to environmental conditions.

Reflections on a Jack Pine Cone

Holding a jack pine cone in a flame and watching it open provides such a good exhibit that I often make use of it when speaking to groups about boreal ecology. Over the years I have held cones in campfires, in laboratory Bunsen burners, and over candles at log cabin lodges. But I remember one workshop above the rest, a session given to a group of northern teachers. I held the cone in a flame; the cone heated and finally ignited into flames. I removed it from the fire, and those around the table watched as the segments separated and the cone blossomed. Shortly, the flames burned out, and I tapped the cone against the surface of the table. After a number of taps several seeds skidded out onto the table, unscorched. I set the branch down and went on to explain several other aspects of the fire ecology of jack pine. At the end of the session, I happened to pick up the branch and haphazardly rapped the cooled cone once more against the table top. Instantly the table was dotted with jack pine seeds. I was speechless—but nothing needed to be said. The point was clear to all: Seeds are not released from a hot cone. Not until the cone (and presumably the area surrounding it) is thoroughly cooled are seeds readily released. In a burnt-over area, it may take several days for the cone to cool. In my demonstration, it took less than an hour.

This experience made me contemplate the evolution of serotiny and wonder what other refinements might be programmed into these fire-sensitive cones. I decided to devise a small experiment to test whether what I had observed with this group was a fluke. I also wanted to observe the whole process more closely. Consequently, one fine March day I collected cones from jack pines of different sizes and ages, let them thaw for several days, then held one after another in a flame. From my notes made during these cone-burning sessions I recorded the following:

> One of the first things to happen is that resin starts to ooze out of the cone at different places on its surface. Soon after that, the resin ignites and engulfs the cone in a soft flame. This is a fairly slow-burning resin, just enough to keep the fire going. It does not seem to build up in heat

and sputter and spit as a hotter-burning substance would. The resin creates a gentle, lamplike flame around the cone. An interesting design feature is that this resin oozes out from a pore located in the center of each hard surface segment of the cone. The main amount of resin is emitted from this pore, it does not ooze out from the edge of the segment in the space that leads to where the seeds are stored.

I cut several cones lengthwise in half with a coping saw and carefully sanded and polished the cut surface. Inside each segment, and connected with the pore, was a ductlike structure, which seemed to carry the resin from the interior of the cone to its surface in response to heat. Thus there seemed to be a structure inside each segment of the cone that released resin once the cone was heated past a critical temperature. The resin then ignited, burnt softly for an average of 93 seconds (between 45 and 177 seconds for the thirty cones tested), and then, the resin having been consumed, this flame extinguished itself. It seemed that, once ignited, the cone was programmed to provide a flame for the right amount of time to open the cone. I was left with the distinct impression that, while a forest fire is needed to initiate this process, the cone itself is capable of providing the type and duration of flame it needs to open and disperse its seeds.

Next I wanted to determine whether what the group and I had watched was a fluke or the norm: After having ignited and opened, does a warm cone retain its seeds while a cool cone releases its seeds? Cones at many different temperatures could have been tested, but I decided to test the two extremes (hot and cool). I simply tapped the cone against a tabletop while it was hot and counted the number of seeds released. I then allowed the cone to cool thoroughly (two hours) before repeating the procedure. Some typical results from my cone-tapping tests:

> One cone was tapped against the table twenty-five times while hot, and no seeds were released. I let the cone cool for two hours. I tapped it four times, and eighteen seeds were released. I tapped another cone twenty-five times and only got three seeds out of it. I let it cool for the two hours and tapped it five times, and twenty-three seeds were released. I tapped another cone twenty-five times, and no seeds were released. I let it cool for the standard two hours; three seeds fell out of it just turning the cone over. After five taps, eighteen seeds were scattered on the paper.

In my sample of thirty cones, two appeared infertile—that is, they did not contain any seeds. The others gave results similar to the ones above. My conclusion? These results should be verified in an actual burnt-over area, and tests should be conducted to see if wind and rain jiggle the cooled cones enough to release the seeds over the course of several days or weeks. Nevertheless, there does appear to be a mechanism that holds jack pine seeds in the cone while it is hot and releases them only after the cone has cooled. One way this may be accomplished involves the fine translucent hairs covering one side of each seed compartment. These may become sticky when warm, thus holding the seeds in place. Once cooled, they may lose their stickiness, allowing the seeds to be released more readily.

These observations and simple experiments make one appreciate the elegance of design that natural selection has created in a jack pine cone. Jack pine trees have been shaped in a number of important ways by the force of fire. The selection pressures imposed on jack pine by frequent exposure to fire have caused jack pine to evolve such that it cannot persevere in the prolonged absence of fire. It has actually evolved into a species that is dependent on forest fires for its survival.

Summary
The adaptive responses of moose and jack pine to forest fires in the North American taiga are clearly of importance to the evolutionary success of these species. Other species in the boreal forest also show impressive adaptations to forest fires, among them black spruce, white or paper birch, fireweed *(Epilobium angustifolium)*, beaked willow *(Salix bebbiana)*, beaver, grizzly bear *(Ursus arctos)*, and three-toed and black-backed woodpeckers *(Picoides tridactylus* and *Picoides arcticus)*.

In the following chapter, instead of exploring how each of these individual species has adapted to forest fires, I change focus and explore how ecological communities respond to forest fires.

General References
Allen, D. L. 1979. *Wolves of Minong: Their Vital Role in a Wild Community*. Boston: Houghton Mifflin.
Cayford, J. H., and D. J. McRae. 1983. The Ecological Role of Fire in Jack Pine Forests. In *The Role of Fire in Northern Circumpolar Ecosystems,* edited by R. W. Wein and D. A. MacLean. New York: John Wiley and Sons.
Franzmann, A. W., and C. C. Schwartz, eds. 1997. *Ecology and Management of the North American Moose*. Washington, D.C.: Smithsonian Institution Press.
Geist, V. 1971. *Mountain Sheep: A Study in Behavior and Evolution*. Chicago: University of Chicago Press.
Henry, J. D. 1996. *How to Spot a Fox*. Shelburne, Vt.: Chapters Publishing.
Krebs, C. J., S. Boutin, and R. Boonstra, eds. 2001. *Ecosystem Dynamics of the Boreal Forest: The Kluane Project*. Oxford: Oxford University Press.
Nelson, R. K. 1983. *Make Prayers to the Raven: A Koyukon View of the Northern Forest*. Chicago: University of Chicago Press.
Rowe, J. S. 1983. Concepts of Fire Effects on Plant Individuals and Species. In *The Role of Fire in Northern Circumpolar Ecosystems,* edited by R. W. Wein and D. A. MacLean. New York: John Wiley and Sons.
Viereck, L. A. 1973. Ecological Effects of River Flooding and Forest Fires on Permafrost in the Taiga of Alaska. In *Permafrost*. Washington, D.C.: National Academy of Science.

6

A Forest in Search of a Fire

Does a forest fire simply consume trees and shrubs, and then the forest starts to recover and grow again? Or has the role of forest fires in the taiga been so extensive and profound that it has affected community dynamics and even the way plant communities develop and mature?

To explore these questions, I use the theory of secondary plant succession to look at the sequence of plant communities as they grow and replace each other and then reflect upon the long association of forest fire with the taiga.

Plant Dynamics in the Boreal Forest

Secondary plant succession is the process by which an ecosystem recovers after a major disturbance, such as a flood or a forest fire. (Primary plant succession—a related ecological concept—is the process by which a plant ecosystem initially establishes itself on recently formed or exposed substrate, such as vegetation developing on newly formed sand dunes or land recently uncovered by the melting of a glacier.) The theory of secondary plant succession—very simply put—states that most natural habitats, be they land or water, show a succession of plant communities after the area has been devastated by a natural catastrophe. The site will first be invaded by opportunistic, or pioneer, species, which colonize a site and begin to create better conditions for growth (e.g., by establishing organic soil, increased shade, or protection from wind or water currents). As ecological conditions improve, the pioneer species gradually gives way to larger plants, and over time a number of species will grow on the site. The pattern of increased productivity and diversity continues through the early and middle stages of succession and communities continues until the climax community (the most stable) is reached.

Some ecologists doubt the existence of stable climax communities in biomes like the boreal forest, noting that major disturbances to these communities are so frequent that a climax community seldom develops or, if it does, does not continue for very long. These critiques notwithstanding, the theory of succession states that an ecological community is in a climax state when it is self-perpetuating—that is, when the vegetation maintains and replaces itself over time, without any major changes in composition.

How well does the theory of secondary succession describe the boreal forest? Researchers have found that, after a catastrophe, successional pathways in the boreal forest are short and uncomplicated. Early succession follows a simple two-step process: First, light-seeded pioneer species (e.g., fireweed, willow, aspen, and white birch) establish themselves. Second, other tree, shrub, and herb species (e.g., white spruce, balsam fir, green alder *[Alnus crispa]*, and wild sarsaparilla *[Aralia nudicaulis]*) sprout or germinate. Usually, two—at most three—successive communities grow on a site before a disturbance such as a forest fire devastates the site again.

Stan Rowe, professor emeritus from the University of Saskatchewan, has studied secondary succession in the boreal forest and finds that boreal plant species show one or more of five adaptive responses to cope with natural disasters like forest fires. He terms the plants according to these responses: invaders, evaders, resisters, endurers, and avoiders.

Invaders—such as fireweed, Canada thistle *(Cirsium arvense)*, beaked willow, flat-leaved willow *(Salix planifolia)*, aspen, and paper birch—are early arrivers that owe their success to their copious production of short-lived, wind-dispersed seeds. Not unexpectedly, they are Sun-tolerant species and do well on exposed mineral soil. Invaders are valuable members of the boreal flora because they quickly add much organic matter to the soils of badly disturbed sites.

Evaders evade the destruction wrought by a disturbance like a forest fire by storing their relatively long-lived seeds in the canopy, the humus, or the mineral soil. Jack pine and lodgepole pine, with their serotinous cones, successfully use the evader tactic. Black spruce develops dense crops of semiserotinous cones in its crown. These heavy clusters of cones create the club-topped black spruce familiar to anyone who has canoed through the taiga. Evader plant species that "bank" their seeds in the humus or mineral soil can be divided into two groups: ephemeral plants and shade-tolerant perennials. Short-lived plants—hedge bindweed *(Calystegia sepium)*, bristly sarsaparilla *(Aralia hispida)*, pink corydalis *(Corydalis sempervirens)*, Bicknell's geranium *(Geranium bicknellii)*—grow quickly after a fire and deposit their seeds all at once. These hard-coated seeds lie dormant until exposed to the heat of the next fire. The second group deposits seed slowly over time. These seeds usually do not germinate until a new forest canopy has been established, and their plants often persist into the late successional stages. This group includes the berry shrubs, such as common snowberry *(Symphoricarpos albus)*, soapberry *(Shepherdia canadensis)*, currant *(Ribes* species), bush cranberry *(Viburnum* species), and red-osier dogwood *(Cornus stolonifera)*, whose seeds are stored at various levels in the humus, often with the assistance of the birds and mammals that feed on the berries and help spread the seeds.

Resisters are plant species that resist fire damage. Jack pine and lodgepole pine, for example, are resisters. Their thick, corky bark gives them some protection against light surface fires, and their habit of shedding dead branches helps lower the frequency of destructive crown fires. These two pine species are the only boreal trees that can be repeatedly scarred by fire and survive and continue to grow. Another resister is sheathed cotton-grass *(Eriophorum vaginatum* ssp *vaginatum)*, which, when mature, resists damage from fire by creating dense, water-saturated tussocks on well-developed wetlands.

Endurers are plant species that send up new shoots after a disturbance. Examples include aspen, paper birch, wild sarsaparilla, spreading dogbane *(Apocynum androsaemifolium)*, bunchberry *(Cornus canadensis)*, and woodland horsetail *(Equisetum sylvaticum)*, to mention just a few. The survival of endurers after a fire or other major disturbance seems to depend greatly on the vertical positioning of the sprouting buds in the insulating humus and mineral soil. If a fire is intense enough to destroy shallowly buried buds or buds around the stem or root crown, then only those species with more deeply buried buds (for example, aspen) will sprout in the postfire community.

Avoiders establish only in late successional stages and often depend upon mature plant communities to provide enough moisture and shade for them to germinate. Balsam fir, white spruce, and twinflower *(Linnea borealis)*, as well as many saprophytic vascular plants such as coralroot *(Corallorhiza* species) and Indian pipe *(Monotropa uniflora)* and certain species of feather moss and slow-growing lichen, are examples of avoiders.

Rebuilding after a Disaster

Given these five strategies, how does succession in the taiga actually proceed after a serious disturbance? If the disturbance—be it fire, windstorm, avalanche, mud slide, or flood—is sufficiently severe, the resisters will be wiped out, and the avoiders will not be seen until late in the successional process. That leaves boreal plants with three strategies to reestablish themselves: invade, evade, or endure. Since fire is the most widespread disturbance in the taiga, I speak mainly in terms of it, although the successional sequence after other disturbances is similar.

Usually the invaders, the light-seeded pioneer species, are the first to arrive. The purples of fireweed and thistle and the new greens of willow, white birch, and aspen appear during the first growing season after the fire. They are particularly likely to germinate if large areas of bare mineral soil have been created by the disaster. Also, immediately after the fire, the seed-banking jack pine and black spruce will deposit an abundance of seeds from their serotinous cones. Often in a postfire area there can be competition between the invaders—the germinated seedlings—and the enduring sprouters. If the fire has not destroyed the mature root system of the previous forest, then the sprouters generally win. Aspen will sucker along its entire root system; birch will sprout around the root collars of its dead trunks. Numerous shrub species, like prickly rose *(Rosa acicularis)*, serviceberry *(Amelanchier alnifolia)*, and high bush-cranberry *(Viburnum edule)*, also sprout. However, if the root systems have been destroyed by the intensity of the fire, then the invaders and evaders will establish the vegetation of the postdisturbance forest.

Thus, depending upon the severity and uniformity of the fire, the site can end up with a forest dominated by aspen, white birch, jack pine, or black spruce—or a combination of these tree species. This range of possibilities reminds us of the complexities of forest fires: different types and intensities of forest fire have different ecological effects and give rise to a variety of postfire forests.

The Indian pipe plant uses the "avoider" adaptive response to forest fires.

The second stage of succession begins after two or three decades. The new forest canopy is well enough established that its shade leads to new plant types germinating on the forest floor. This stage is marked by the slow establishment of avoiders. Furthermore, the composition of the forest begins to change. Aspen, white birch, and jack and lodgepole pine do not germinate in shade conditions; hence these pioneers may gradually be replaced by species able to germinate under an established forest canopy. Black spruce, white spruce, and—in the eastern boreal forest—balsam fir have this ability. However, succession to these shade-tolerant conifers can be a slow process. The supply of conifer seed is often small, distribution of seed over a large area can be slow, and seedbed conditions may inhibit germination. For example, a thick layer of birch or aspen litter can smother and kill newly germinated conifer seedlings. Leslie Viereck, a boreal ecologist who has studied the taiga in Alaska for decades, and others speculate that one benefit of aspen and white birch shedding their leaves every autumn is to partially suppress germination of competing conifers. If this hypothesis is correct, it gives quite a different perspective on the display of fall colors.

Conifer seedlings do eventually manage to establish themselves; one of the most commonly encountered forest stands in the taiga is mature aspen, birch, or jack pine, with shade-tolerant conifers (white spruce, black spruce, balsam fir) of various ages and sizes growing up into the canopy. If the canopy remains undisturbed long enough, the shade-tolerant conifer species will gradually take over the stand, and the composition of the forest will change from pioneering tree species to self-perpetuating, late-successional tree species.

The Forest That Sphagnum Builds

Because of the high frequency with which disturbances, mainly forest fires, devastate the taiga, these white spruce, black spruce, and balsam fir forests are not often given a chance to mature before they are disrupted again. However, on some sites, these avoider tree species do mature, and the kind of forest that develops next is greatly influenced by the presence or absence of an obscure group of plants, the mosses. Yes, that's right; in the North a three-inch-high moss greatly influences how a mature taiga forest develops. As Robert Service wrote, "There are strange things done in the midnight sun." It is as true for northern ecology as it is for people moiling for gold. If a forest stand in the taiga is moist enough and cool enough to allow a carpet of sphagnum moss or feather moss to become established under the canopy, then these small mosses will gradually but dramatically alter the whole composition of this forest.

To understand the considerable power wielded by these boreal mosses, let us first look at the type of late successional forest that is likely to develop if mosses do *not* become established. In the southern boreal forest and in drier, warmer sites farther north, lack of moisture prevents a moss carpet from being established. Furthermore, on islands and lands adjacent to rivers and creeks, periodic flooding may disrupt a moss layer from developing. On these sites, white spruce and balsam fir may grow to become magnificent old growth stands, forests as old as 350 years and trees with trunks four feet (1.2 m) or more in diameter. However, forests are usually damaged—by fire, wind, insect infestation, or humans with chainsaws—before they attain such dimensions. Old growth forests are protected in a few Canadian national parks, such as Prince Albert and Wood Buffalo.

On sites moist and cool enough, a carpet of feather moss or sphagnum moss forms. The three most common feather mosses in the Canadian taiga are red-stemmed feather moss *(Pleurozium schreberi)*, stair-step moss *(Hylocomium splendens)*, and knight's plume *(Ptilium crista-castrensis)*. Other species of feather moss are present but less common. In the Canadian taiga there are more than forty species of moss, all of them belonging to the genus *Sphagnum*. Different conditions promote the establishment of different mosses—either feather moss or sphagnum moss. If moisture is sufficient, and direct sunlight reaches the forest floor, a species of sphagnum moss is likely to become established. A little less moisture and shadier conditions favor feather moss.

The sphagnum mosses are about as tall as my thumb, yet as a group they are plants of considerable potency. Through a process called paludification, or swamping, they ultimately determine the structure and composition of the mature taiga forest. A continuous carpet of sphagnum moss is like a huge, moist, cold sponge. Sphagnum moss becomes water saturated, and it blocks any drainage of water to and from the site. Sphagnum also produces considerable acid, enough to suppress the activity of bacteria and other microorganisms that carry out decomposition. Once this important ecological process is reduced or inhibited, dead organic matter, particularly dead sphagnum moss, accumulates in the form of a thick layer of peat moss. The living sphagnum moss continues to acidify its surroundings, until many of the plants that formerly occupied the site die out and only acid-loving plants— such as black spruce, bog rosemary *(Andromeda polifolia)*, Labrador tea *(Ledum groenlandicum)*, leatherleaf *(Chamaedaphne calyculata)*, and swamp birch *(Betula glandulosa)*—are able to grow. Organic deposits collect further, until the surface of the ground is occupied by a landform composed entirely of accumulated peat. (Paludification and the development of organic terrain are examined in greater detail in Chapter 9.)

If moisture conditions on the site are more restricted and if there is a closed canopy preventing sunlight from reaching the forest floor, feather moss, rather than sphagnum moss, is more likely to develop. Feather moss initiates a series of events similar to, but not as dramatic as, that of sphagnum moss. For example, feather moss is only slightly acidic, enough to favor the establishment of black spruce but not enough to completely suppress decomposition. Consequently, accumulations of feather moss peat are not nearly as thick as those of sphagnum peat. However, the layer of live-on-dead feather moss that does form is insulation enough that the temperature of the soil is dramatically reduced. Decomposition is suppressed, and permafrost begins to form.

Cold soils also favor the establishment of black spruce over other boreal trees. Slowly the stand changes into a forest of black spruce, with a dense closed canopy and a continuous green carpet of feather moss. Lack of sunlight and mildly acidic conditions eliminate many herbs and shrub species, the exceptions being Labrador tea and a few other acid-loving plants, which grow under openings in the canopy. Black spruce–feather moss stands are so distinctive in the southern boreal forest that they have been nicknamed spruce moss yards. The name captures the ambience of the stand: a dense growth of spruce, a carpet of feather moss, a few herbs and shrubs, and little else. If moisture begins to accumulate in the spruce moss yard and a few trees topple, allowing sunlight to reach the forest floor, sphagnum moss may become established, perhaps changing the black spruce–feather moss forest into a black spruce–sphagnum moss forest.

Aging Ungracefully: Self-Perpetuating Taiga Forests

The two types of black spruce forest, with their highly simplified composition, are considered by some ecologists to be the climax vegetation of the Canadian taiga. Black spruce is able to reproduce under its own shade, and the acid, cold soils help perpetuate the vegetation of these two depauperate forests. As a result, unless a drastic disturbance occurs, they are largely self-perpetuating—the characteristic feature of climax vegetation. However, as end points of succession, black spruce forests are not exactly thriving. Black spruce does not thin itself well; that is, the weaker trees do not die off, making room for the more vigorous trees to grow larger. In a spruce moss yard, the trees are fairly small, and there are too many of them for good growth. The annual growth rings of these black spruce confirms that growth is suppressed; the outermost rings are often only a fraction of a millimeter wide.

In the more open black spruce–sphagnum forest there appears to be plenty of room for growth, and abundant sunlight does reach ground level, but tree growth is inhibited by the cold, acidic, nutrient-poor sphagnum peat soil. Some northern plant species have evolved special tactics to cope with these soils. In Alaska, K. Kielland shows that some northern plants, with assistance from mycorrhizal fungi, are able to absorb whole amino acids into their roots as an adaptation to the low inorganic nitrogen content of northern soils.

Trees growing on sphagnum peat can be extremely suppressed: gnarled black spruce or tamarack that stand only twenty-five feet (8 m) tall and have trunks only three inches (8 cm) in diameter are often found to be over a hundred years old—not much to show for a century of growth. Without vigorous plant growth, foraging opportunities for herbivores are limited. Without abundant and diverse herbivores, predators do not thrive. Consequently, wildlife diversity is often very low.

Why are the climax communities of the taiga so biologically unproductive, and why do they support so few animals? In climax communities of other biomes, it is not unusual for late successional plants to show a lower rate of photosynthesis than early successional plants. In these climax communities, the soils are often less fertile because dead plant material (e.g., trunks and branches of trees) have accumulated, locking up many of the nutrients. Yet in these biomes (e.g., eastern deciduous forest, temperate rainforest), the late successional communities are structurally more complex and often show higher species diversity; some species, like the spotted owl *(Strix occidentalis)* of the Pacific Northwest, even specialize in these old-growth forests. None of these attributes appears to be true of the taiga's climax communities. Primary productivity, decomposition, and species diversity are all reduced. In addition, over much of the North American taiga, as forest stands age they develop a cluttered appearance. This is true for the taiga's climax communities, but it is equally true for earlier successional forests of the taiga, such as aspen and birch stands.

In mature aspen forests, dead branches, which are slow to decompose, litter the forest floor. Clusters of diamond willow or green alder shoots, dead for many years but still standing, are scattered throughout aspen and birch forests, like giant sheaves of wheat long ago abandoned. White birch trees as they mature become swathed in papery bark, which hangs in tatters from their trunks. In black spruce–feather moss forests, often only the tops of the trees have living branches; their trunks are covered with dead branches and twigs, many of which are blanketed by arboreal lichens. These dead branches can be so dense that the stands are nearly impenetrable. Often dead spruce are held upright by the network of dead branches.

Not all forests become shaggy and unkempt as they mature. Beech, oak, eastern white pine, and hemlock grow massive trunks that form a dense, closed canopy and create an almost cathedral-like space underneath. The forest floor in these mixed hardwood stands is deep in leaves, with a few shade-tolerant shrubs adding a touch of greenery. Such a manicured look is far from the disheveled appearance of the mature boreal forest.

One factor explaining the cluttered appearance of the mature taiga is that it is a detritus ecosystem, in which dead plant material accumulates because decomposition occurs so slowly. Six or seven months of winter, cool nights during the remainder of the year, permafrost soils, and the acidification carried out by sphagnum moss contribute to the reduced rate of decomposition. Certain tundra ecosystems, particularly those near treeline but even some in the high Arctic, are similar—they are also detritus ecosystems, in which plant growth exceeds decomposition. In such detritus ecosystems, one of the most important agents of decomposition is fire. Thus for much of the North American taiga fire is not only the dominant agent of disturbance but also the dominant agent of decomposition.

Promoting Forest Renewal

While the taiga is clearly a detritus ecosystem, there is more to the cluttered appearance of the mature taiga than this. Let's look at it from an evolutionary perspective. The theory of evolution applies equally well to plant and animal species. When we apply it to the evolution of trees, it dictates that the adaptive strategies that each tree species has evolved will serve the biological self-interest of that tree. This means that each tree in the forest should be genetically programmed to leave as many offspring in the next generation as possible. Maximizing reproductive fitness, whether in a plant or an animal, is viewed as the driving force of evolution. Consequently, everything else being equal, trees are probably selected not only for any feature that helps them to produce a maximum number of seeds but also for any mechanism that releases these seeds in a manner and at a time when their offspring have the best chance to germinate and prosper.

Given these dynamics and given the unproductive quality of the taiga's late successional forests, certain tree species of the taiga may evolve characteristics that resist the development of the climax community and may instead promote forest regeneration under conditions favorable to their offspring. Another way of understanding this point is to ask the question: Has natural selection slowly developed features in at least some of these plant species that tend to keep the taiga in an earlier, more productive stage of succession? Perhaps certain tree species have not only developed adaptations by which they can successfully reestablish themselves after a fire but have also taken the next step and evolved characteristics that actively promote the occurrence of a forest fire. Perhaps the cluttered appearance of the mature taiga is an expression of the inherent flammable characteristics that develop as certain taiga trees age.

Robert Mutch, of the U.S. Forest Service, is one of the main proponents of this interesting, yet controversial, school of thought. He suggests that plant communities rejuvenated by fire may burn more readily and more frequently than plant communities that are not fire dependent. In certain plant species, natural selection seems to favor the development of characteristics that make these plants more flammable as they age. Mutch believes that a number of fire-dependent ecosystems show these patterns, such as ponderosa pine forests, savanna woodlands, and arid grassland communities, and that fire-promoting characteristics are also strongly evident in much of the circumpolar taiga.

Mutch's hypothesis has been criticized by some ecologists. Its controversial dimension revolves around the contention that natural selection has directly shaped the aging process of taiga trees and shrubs as well as the way they decompose so that the chances of a forest fire starting and being supported are enhanced. They ask: Do dead branches actually decompose in certain ways so as to develop flammable characteristics? Do dead leaves as they decompose accumulate easily ignitable oils as a tactic for increasing the chances of a forest fire? To many scientists, Mutch's theory seems far-fetched, and they reject it. And yet to me, the hypothesis gives potentially interesting insights. Why do white and black spruce develop into "torch" trees as they age? Other trees shed their dead branches, but black and white spruce seem to hoard them. White birch as it grows breaks up its papery bark, but it is not shed; instead it hangs from the tree forming a highly flammable feature. Other tree species in other forests exfoliate and completely shed their dead bark—why not white birch?

Mutch's hypothesis is also useful because it makes us think more carefully about the factors that actually sustain a forest fire in the taiga. The traditional viewpoint is that weather and the amount of deadfall and other fire fuels in a stand determine the fire hazard of that forest. More specifically, the moisture content of the fuels in the stand is usually considered the most important factor for predicting the hazard of a forest fire. However, Mutch suggests that there is more to it than that. He speculates that the physical structure of the vegetative community as well as the evolved chemical properties of the litter and other fire fuels found in that forest may to a great extent determine the spread and intensity of a fire. He maintains that properties develop in the forest's dead twigs, branches, and forest floor litter so that, once ignited, they release their energy at a rapid rate and help to sustain the fire. In short, Mutch suggests that the structure of the taiga's vegetation sets the stage for the fire, while the weather simply determines the time of its performance. If his hypothesis is correct, the drama of a forest fire may to a great extent be determined and orchestrated by the vegetation itself. It is a challenging and thought-provoking hypothesis, one that leads to interesting insights about the taiga and that should be the focus of further research and experimentation.

Summary

For much of the North American taiga, forest fires are not only the main agent of disturbance, creating a mosaic of habitats and elevating species diversity, but also a major agent of decomposition. Moose, jack pine, and a number of other plant and animal species in the taiga have evolved adaptive strategies in response to forest fires. Furthermore, it may be that, as many plant communities in the boreal forest age, they actually promote the occurrence of forest fires. For all of these reasons, the North American taiga should be viewed as a fire-dependent forest.

General References

Dix, R. L., and J. M. A. Swan. 1971. The Role of Disturbance and Succession in Upland Forest at Candle Lake, Saskatchewan. *Canadian Journal of Botany* 49:657–76.

Heilman, P. E. 1966. Change in Distribution and Availability of Nitrogen with Forest Succession on North Slopes in Interior Alaska. *Ecology* 47:825–31.

Heinselman, M. L. 1973. Fire in the Virgin Forests of the Boundary Waters Canoe Area, Minnesota. *Quaternary Research* 3:329–82.

Kielland, K. Role of Free Amino Acids in the Nitrogen Economy of Arctic Cryptogams. *Ecoscience* 4:75–79

Romme, W. H., and D. G. Despain. 1989. The Yellowstone Fires. *Scientific American* 261:37–46.

Service, R. 1983. *The Collected Poems of Robert Service*. Toronto: McGraw-Hill Ryerson.

Van Cleve, K., F. S. Chapin III, P. W. Flanagan, L. A. Viereck, and C. T. Dyrness, eds. 1986. *Forest Ecosystems in the Alaskan Taiga: A Synthesis of Structure and Function*. New York: Springer-Verlag.

Van Cleve, K., C. T. Dyrness, L. A. Viereck, J. Fox, F. S. Chapin III, and W. Oechel. 1983. Taiga Ecosystems in Interior Alaska. *BioScience* 33:39–44.

Wein, R. W., R. R. Riewe, and I. R. Methven. 1983. *Resources and Dynamics of the Boreal Zone*. Ottawa: Association of Canadian Universities for Northern Studies.

7
The Taiga in Winter

A wildlife biologist on snowshoes plods across a remote northern lake pulling a toboggan. The temperature is –50°F (–45°C). A north wind is blowing, adding a strong wind-chill factor to the bitter cold. Bundled in a felt face mask, parka, wind pants, mukluks, and moose-hide gauntlets, the biologist shuffles along, using ski poles to help haul the toboggan. He treks across the lake and into an isolated bay on its far side, where he stops in front of a snow-covered hump. Within moments the biologist has slipped out of the lampwick snowshoe harnesses and is kneeling down, untying a long metal rod from the toboggan. Gingerly, so as to leave most of the snow undisturbed, he walks over to the hump, a deserted beaver lodge, and thrusts the metal rod through the snow and into the wall of the lodge. Pushing and prying, he jams it farther and farther in until he can feel the point strike the ice floor. Slowly he removes the rod, leaving a small, neat tunnel into the heart of this abandoned dwelling.

Returning to his toboggan, he unties a flexible wooden stick with a thermometer attached to the end. Carefully, he lowers this device through the tunnel into the center of the lodge. Then, as painstakingly as a painter finishing a canvas, he brushes and smoothes the snow until there is nearly an airtight seal around the stick, the end of which protrudes from the lodge like a huge dissecting pin. The biologist slips into his snowshoes again and begins to circle the lodge, about twenty yards (18 m) out from it, looking for the tracks of any animal—weasel, mink, otter, or mouse—that may be occupying the lodge. Finding none after circling twice, he sets off along the lakeshore in an effort to stay warm in this bitter cold. One of his ski poles is marked off in centimeters, and he pokes it into the soft taiga snow a number of times until he decides that on average approximately nine inches (22 cm) of snow covers the ice and surroundings of this sheltered bay. Twenty minutes later he returns to the beaver lodge, extracts the stick and thermometer, and reads the temperature. In a small, red, field notebook, he quickly records his findings: "beaver lodge #10; –1°C."

The temperature at the center of the lodge is only a degree below the freezing temperature of water and a full 80°F (44°C) warmer than the temperature of the air outside the lodge. That difference is equivalent to the difference between a normal winter's day and quite a warm summer's day; it is a temperature range that spans almost half the distance between the freezing point and the boiling point of water.

The biologist has been measuring one of the hot spots of the boreal forest, a relatively warm microhabitat in the frozen landscape. Had he not intentionally chosen an unoccupied beaver lodge, he might have assumed that the temperature inside the lodge was elevated by the body heat of its inhabitants. In fact, the absence of tracks suggests that there have been no animals using the lodge recently. This freakishly warm site is caused by an interaction of the cold and snow, and an understanding of that important relationship is central to understanding the winter ecology of the taiga.

By anyone's calculation winter in the taiga is fierce. Snow covers the ground for six or seven months of the year. The temperature can go down to −50°F (−45°C) and stay at that temperature day and night for several weeks. Sunlight is at a premium. In this area of northern Saskatchewan where the biologist is working, near the winter solstice the Sun rises at 10:00 A.M and sets by 4:00 P.M. Farther north those six hours of sunlight would seem like a luxury. As described in Chapter 2, from late October until early March these northern regions sink into a negative heat balance: More energy is lost to the atmosphere and outer space than the solar energy gained. The farther north, the greater the deficit. It is, to say the least, disconcerting to be a warm-blooded creature trying to exist in a world that operates on a net energy loss for over a third of the year. No wonder northerners have a complex psychological response to winter!

The boreal forest is often colder than the tundra, even though the latter is more exposed to winds. The oceans close to the Arctic tundra influence its weather, bringing it not only bitterly cold wind but also the warming effect of the ocean. Thus the coldest air temperatures for both North America and Eurasia have been recorded in taiga environments: During January 1999, some locations near Fairbanks, Alaska, recorded −94°F (−70°C), and two decades earlier the town of Verkhoyansk at the foot of the Cherskiy Mountains in northeastern Siberia recorded −103°F (−75°C).

Snow often seems to be the only part of the boreal forest landscape not frozen rock hard at 40 below zero. Soft and fluffy snow is characteristic of a taiga winter. A pail overflowing with taiga snow brought into a cabin to melt will yield only an inch of water in the bottom of the pail. Taiga snow is like spun gossamer, different from the snow that breaks limbs off trees in the East and the Midwest and different also from the snow that causes avalanches in western mountains. It is a special substance, the quintessence of snow, something that approaches crystallized vapor.

Snowflakes: Boreal Crystals

Northerners usually have a positive response to the coming of snow, often remarking that a good covering of snow makes things warmer. There appears to be some truth to that. For example, for beavers in their lodges, one of the most difficult time of winter is after the lakes have frozen and before the snow has fallen. To understand the warmth that snow brings, we must explore the formation of a single snowflake.

Most people assume that a droplet of water freezes into ice whenever its temperature falls below 32°F (0°C). Research shows, however, that it is far from that simple. Microscopic droplets of highly purified water must be taken to temperatures of −40°F (which is also −40°C) before they begin to freeze. It is clear that something solid must be in the water for crystals or snowflakes to form. Highly purified water has no solids—no impurities—and thus has no "seeds" for the water molecules to bond onto; water in that condition can be supercooled without freezing. The impurities in water—tiny dust and clay particles, microscopic fragments of decaying plants, even bacteria—are the seeds (the scientific term is *nucleating agents*) that allow water molecules to change from a liquid to a solid.

For years meteorologists thought that dust and clay particles in the atmosphere were the most common nucleating agents for the formation of snowflakes. When researchers quantified snowflakes forming in clouds, however, they found a thousand times more snowflakes than dust and clay nucleating agents. This discrepancy has led to a number of hypotheses concerning how snowflakes form. One of the most interesting theories has been put forward by Russell Schnell, of the University of Colorado, who says that "at warmer (cloud) temperature, all snow is probably biologically initiated." He believes that two common, almost ubiquitous, species of bacteria—*Pseudomanas sydngae* and *Erwina herbicola*—give off a waste molecule with the greatest affinity for water ever measured. The strong bonding force of water molecules onto the surface of these bacteria promotes the formation of snow. Schnell has demonstrated this relationship experimentally: When he injects these bacteria into a chilled, humid cloud chamber, snow immediately forms. If the chamber is warmer, raindrops form. When Schnell dusts these bacteria onto the surface of cooled plants, frost forms. His and other researchers' calculations show that these two bacteria, either blown off plants or up from ocean waves, account for the formation of much of the world's snow and rain.

Snowflakes may also form from a process called splintering. John Halkett, a physicist at the University of Nevada, uses an experience familiar to everyone to explain splintering: "When you take a glass of water and put it in the freezer, you will end up with a broken glass." It is the expansion of water as it freezes that breaks the glass. What causes this expansion? When water molecules link to form a crystal, they take up more room than they do as liquid molecules sliding and moving around each other. In real clouds, each droplet of water may act like a glass: When the droplet freezes on the outside, it traps liquid water at the center, which eventually freezes. As this interior water changes into solid ice, the droplet explodes, and splinters of the ice crystals can scatter, becoming nucleating agents for the formation of snowflakes.

The beautiful branched form of a snowflake is determined by the geometry of the water molecule. Its polarity and shape determine how water molecules bond together, and these properties naturally give the snowflake its hexagonal shape. More specifically, water molecules can bond together in only two directions, vertically or horizontally, and the direction of the bonding is determined largely by temperature. (Other conditions such as wind can change the shape of the crystal once it is formed.) The stellar, or dendrite, pattern (large, intricate, six-pointed snowflakes) is only one of the shapes a snowflake can assume (the table outlines the various shapes that a snowflake takes on, according to the temperature at which it forms). Once a snowflake forms, its shape can be broken or modified by the wind; humidity can build up a rime on it; or the snowflake can fall through air of different temperatures, which modifies its shape.

When snow falls in the boreal forest, the temperature is typically between –4° and 15°F (–20° and –10°C), and often there is little wind. These moderately cold and calm conditions promote the formation of large and intricate snowflakes and preserve

Type of Snow Crystal

Forms	Temperature Range for Formation	
Thin hexagonal plates and stars	32 to 27°F	(0 to –3°C)
Needles	27 to 23°F	(–3 to –5°C)
Prismatic hollow plates	23 to 18°F	(–5 to –8°C)
Medium hexagonal plates	18 to 10°F	(–8 to –12°C)
Stellars and dendrites	10 to 3°F	(–12 to –16°C)
Thick hexagonal plates	3 to –13°F	(–16 to –25°C)
Hollow prisms	–13 to –58°F	(–25 to –50°C)

their shape during their passage to Earth. That is, the snowflakes do not metamorphose into ice granules, hail, snow pellets, or any other altered state as they fall. Therein lies the true uniqueness of taiga snow. In the boreal forest the snow on the ground is characteristically soft and fluffy because it tends to be made of feathery, dendritic, snow crystals. Not only are the individual flakes filled with tiny air spaces, but also, as those flakes accumulate on the ground, the tiny air spaces between flakes multiply exponentially.

The Warmth of Snow

How is it that taiga snow can keep us warm? Let's compare a patch of this snow to Fiberglass insulation. Contrary to what many may believe, it is not the Fiberglass itself, but the innumerable tiny air spaces within it, that keeps your house warm. The air in these spaces is perfectly still. And this is the structure of any good insulating material, be it the goose down in an Arctic parka, the tiny hollow beads forming a Styrofoam cup, or the soft, fluffy layers of taiga snow, and it works like this: Air circulating between a hot surface and a cold surface sets up a convection current, which transfers warmth from the warm surface to the cold surface. Moist moving air is twenty times better at conducting heat than still dry air. However, if an insulating material contains many dead air spaces, and if that air is relatively dry, that material constitutes superb insulation. Taiga snow has these characteristics.

Snow is not always an effective insulator. The snowflakes of the tundra, buffeted by the wind, collide with each other and become needlelike slivers rather than dendrite crystals. As these snow needles are rolled by the wind, they tend to line up with each other, and the wind packs them on the ground like toothpicks in a box. As the wind continues to pack these snow needles, tundra snow become nearly as hard as concrete. Wind-packed tundra snow has an insulation value approximately 20 percent that of taiga snow. Tundra snow and taiga snow are thus very different entities, and each produces its own characteristic ecological effects.

Snow even when it is lying in a protected forest environment is a dynamic substance, responding to small changes in temperature or humidity. Peter Marchand describes three metamorphic processes that snow crystals undergo as they are influenced by conditions within the snow pack and by external weather. Destructive, or equitemperature, metamorphism occurs when the H_2O molecules of snowflakes are transferred from one snowflake to another because of small differences in temperature or humidity, with a resulting loss of the snow crystals' fine structure. These snowflakes become rounded ice grains interconnected by thin necks. Constructive, or temperature-gradient, metamorphism occurs when a strong temperature gradient is present in the snow cover—colder at the top surface and warmer near the ground. Because water vapor in suspension increases with increases in temperature, water vapor moves from the lower layers to the upper layers. The formation of sugar snow, or depth hoar, next to the ground results from this process. Melt metamorphism occurs during late winter and spring. Surface meltwater percolates downward and refreezes in the colder layers, dramatically altering the structure of the snow cover.

However, let's consider taiga snow during the prolonged cold of a boreal winter. When the snow lies thick and even in the forest, it insulates the ground well. Biologists have found that when fluffy taiga snow on the forest floor reaches the critical thickness of six to eight inches (15 to 20 cm), the temperature next to the ground fluctuates little, even though the air temperature can fluctuate as much as 30° to 50°F (15° to 30°C) from day to day. Also, any summer heat that remains in the ground, as well as any geothermal heat moving up through the crust of the Earth, is trapped next to the ground by this insulating blanket of snow. Consequently, the subnivean temperature next to the ground for most of the winter is often no colder than 25° to 28°F (−4° to −2°C), even when the outside temperature plummets to −40°.

The same is true of lakes and ponds when they are covered by six to eight inches of soft, fluffy snow. The biologist probing the center of the beaver lodge was verifying this. Specifically, he was documenting that when covered by a good blanket of snow the residual heat of the lake warms the lodge, so that for most of the winter its interior is only slightly below the freezing point of water. Had the lodge been occupied, the body heat of the beavers could have raised the interior temperature to as much as 43°F (6°C). All that would change, however, if a strong wind came along, sweeping the lake and the lodge clear of its insulating snow—or during March and April, when the heat of the Sun melts the snow off the lodge. Once the lodge loses its insulating cover of snow, its internal temperature slowly sinks to whatever the outside air temperature is. Consequently, a family of beavers inside a lodge is often more cold stressed during early or late winter or after a windstorm than during the coldest parts of a taiga winter.

Snow Ecology

William O. Pruitt Jr., retired professor of zoology at the University of Manitoba, is one of the pioneers of tundra and taiga snow ecology. He discovered—and adopted—the elaborate aboriginal terminology for the various snow conditions and for their metamorphoses into other conditions. Most of his terms are from the language of the Kovakmiut, the Forest Eskimos of northwestern Alaska, whose vocabulary includes thirty words for snow.

The snow that accumulates on the ground, the typical soft and fluffy snow found in the boreal forest, is *api* (pronounced ah-PEE). The hard, wind-blown snow of the tundra is *upsik* (OOP-sik). The hoar frost that forms on trunks and branches of trees is *kanik* (KAH-nik). And snow that builds up on branches and accumulates in conifer trees is *qali* (KAH-lee). The Kovakmiut also have terms to describe the forms and drifts that tundra snow assumes as it is worked and reworked by the wind.

Pruitt and his colleagues show that qali plays an important role in renewing the forest. In regions of heavy snowfall, qali can build up in a conifer until its weight snaps branches or breaks the main trunk of the tree. Small openings in the forest canopy are thus created, allowing sunlight to reach the forest floor in these areas. The following spring a new growth of Sun-loving plants takes root under these openings, providing food for wildlife. Furthermore, the branches near the top of trees that are heavily laden with qali and kanik can break off in a strong wind, dropping to the ground with their nutritious winter buds, a windfall of winter food for the herbivores of the forest. Animals as diverse as white-tailed deer, snowshoe hares, deer mice, and red-backed voles use these fallen twigs and branches as winter food.

The Kovakmiut word *pukak* (pronounced POO-kak) refers to depth hoar—the relatively open layer of snow that lies next to the ground, under an insulating cover of snow. The residual heat from the ground transforms the bottom inch or two (2 to 5 cm) of snow, so that it sublimes, forming small spaces. More specifically, during this thermal metamorphism, individual snowflakes are changed into pukak crystals. As a result, the inch or two next to the ground is filled with cup-shaped crystals. Small mammals, insects, and spiders move around in this space. C. W. Aitchison has found as many as nineteen species of spider and several species of insect in the pukak, and three species of spiders have even been found to breed there.

In the mountainous areas of the boreal forest, because snow normally falls during the autumn before the ground has cooled and because the snow cover is usually thick, the pukak layer is sometimes six to ten inches (15 to 25 cm) deep. This deep pukak, or sugar snow, is made of distinctive stepped ice crystals. These six-sided scrolls, or cup crystals, form as water vapor slowly sublimes off snow granules low in the snow cover then rises and refreezes on snow granules higher up. A layer of these pukak crystals is fragile and unstable, providing a sliding surface for the snow cover above it. Winter travelers in mountainous wilderness areas should make a practice of driving a ski pole through the snow cover to check for pukak. Do this many times a day. And take the time to dig a snow pit and study the entire profile of the snow pack for the weak layers that promote avalanches.

A good covering of insulating snow is essential for the survival of small mammals during a taiga winter. Bill Pruitt documents that at least six inches (15 cm) of api and a well-developed pukak layer are necessary if mice, voles, shrews, and other small mammals are to survive and prosper during a typical taiga winter. Through field studies, he shows that small mammals have a very low probability of surviving without six to eight inches (15 to 20 cm) of soft insulating snow. In fact, if cold temperatures arrive before the required snow has accumulated on the forest floor there can be a major die-off in small-mammal populations. Furthermore, if small-mammal populations are devastated, owls, weasels, marten, fisher, and foxes (to mention just a few predators) will have difficulty finding sufficient winter food. If the critical thickness of snow accumulates on the forest floor early, however, an important threshold will have been reached, and small mammals will have adequate insulation for their winter survival. Pruitt considers that the accumulation of the required amount of soft taiga snow constitutes the hiemal threshold and, ecologically speaking, marks the official start of winter in the boreal forest. Another crucial period occurs in the spring, when, before the end of severe cold spells, the snow is transformed into crusts and ice particles, thereby losing most of its insulation value. At this time, small mammals again face stress and may die off.

The formation of an icy crust on the surface of the snow—from warm spells, rain, or the warm Sun of late winter—endangers small mammals in their pukak world in another way. These icy crusts can make the snow impermeable to the passage of air and, under certain conditions, cause carbon dioxide (CO_2), from the respiration of plants and animals and from decomposition, to accumulate under the snow. Cheryl Penny, working with Bill Pruitt, has documented the buildup of CO_2 in certain habitats under a snow crust and shows that small mammals actively avoid areas where this gas has accumulated.

Small mammals sometimes respond to icy crusts by digging shafts up through the snow and by gnawing through the crust. I have seen these shafts on the surface of the snow two or three days after a thick ice crust has formed. Some ecologists believe that these shafts allow the CO_2 to exit and fresh air to circulate down into the subnivean environment. Other ecologists consider this explanation part of the folklore of winter ecology. More field studies like Cheryl Penny's will help to settle the debate. Whatever their role, the shafts are a fascinating field sign to watch for, and if you look for them, you won't be alone. I have observed northern owls, such as a great gray owl or northern hawk owl, perch in a tree patiently watching the entrance of one of these shafts—waiting for a mouse or vole to appear.

Larger boreal animals take advantage of the insulating properties of snow in different ways. Snowshoe hares, for example, sleep in the natural snow tents that form in the forest. These are tree trunks or branches so covered in qali that an enclosed chamber is formed. The grouse and ptarmigan are well known for their habit of burrowing under the snow and roosting in subnivean sleeping chambers. Interestingly, as they burrow, they often put into the tunnel some sort of twist or turn before they settle down. I believe the curve in the tunnel is an antipredator device. A coyote, red fox, or other predator may detect the bird from its tracks or the odor at the tunnel's entrance but cannot accurately determine where to pounce because of this curve. If the predator does not pounce directly on the prey, the bird will burst out of the snow and may have a second in which to escape.

Red squirrels are the borderline species, large enough to exist in the ambient air of the taiga—up to a point. Mammals smaller than the red squirrel must live mainly in the under-the-snow pukak to survive the winter; red squirrels can live in the open until the temperature falls below –22°F (–30°C), when they go into underground tunnels. When red squirrels disappear from their arboreal perches and their chatterings no longer form the street talk of the winter woods, then trappers and cross-country skiers alike know that the ambient temperature has fallen below –22°F (–30°C). Snowshoe hares use a similar tactic, leaving the open air to rest in snow tents when it gets intensely cold.

The behavior of chickadees, redpolls, crossbills, and other perching birds poses more questions than we have answers for, but we know a few ways in which these birds make use of the insulating properties of snow. Redpolls, chickadees, snow buntings, and certain sparrows have all been observed to dive into the snow to reduce heat loss on frigid nights. These passerines show other energy-conserving behaviors as well. By roosting for the night under branches heavily laden with qali, these small birds cut down on radiant heat loss, which is especially severe on cold, clear nights. The higher insulating value of feathers compared to fur helps. Another important adaptation is winter hypothermia. S. C. Kendeigh and C. R. Blem discovered that, on a cold night or during other periods of inactivity, chickadees, redpolls, and other bird species allow their body temperature to drop 10° to 12°F (5° to 7°C), resulting in an average savings of 20 percent of the metabolic energy needed to survive a cold winter's night. It's the avian equivalent of turning down the thermostat at night to save on the heating bill.

Certain species of shrews have also evolved a complex response to winter. In his book *The Great Northern Kingdom,* Wayne Lynch states:

> The boreal shrews have come up with a truly novel way to endure winter under the snow—they literally shrivel up. In preparation for winter, a shrew may lose up to 45 percent of its body weight. It doesn't simply lose body fat. The animal's skull shrinks, its backbone shortens, its muscles thin out, and its liver, kidneys, and spleen become smaller. Most remarkable of all is that the animal's brain shrivels by up to a third. By reducing the weight of these vital tissues, the shrew greatly reduces its daily energy needs and improves the chances it has of surviving until spring, earning the one opportunity it will ever have to pass its genes to the next generation.

I believe that future research will uncover other fascinating adaptations by which small birds and mammals cope with the taiga cold.

Snowshoes, Stilts, and Other Animal Inventions

Anyone who has had to plough very far through waist-deep snow knows how difficult and energy demanding it is. It is not surprising that animals have evolved adaptations and developed behaviors to cope with such snow. Several species that inhabit the taiga—wolves, lynx, caribou, and moose—have long legs, high chests, and large feet.

In the soft snow of the taiga, large feet or hooves create a snowshoe effect and help to keep the animal from sinking deeply into the snow. Snowshoe hares, who certainly deserve their name, can spread their toes so much that their feet are as wide as those of a wolf. Ptarmigan and, to a lesser extent, ruffed grouse and spruce grouse grow feathers around the edge of their toes, making their feet into winter snowshoes. The moose has proportionately longer legs than any other member of the deer family. It walks around on virtual stilts. Because of its heavy weight, the moose often sinks right to the ground with every step through soft taiga snow, but its long legs, combined with joints that allow it to lift its legs high, equip the moose for moving through soft and thick taiga snow.

Some species cope with taiga snow by being selective about the snow surfaces upon which they travel. In the early 1980s I carried out a study of red foxes in Prince Albert National Park, in the boreal forest region of Saskatchewan. The study consisted of tracking these foxes during winter over fifty miles (80 km), measuring the various snow surfaces they traveled over while they moved about on their territories. Red foxes, I found, are very selective about which snow surfaces they use for their wintertime travels. They chose areas with a wind crust along creeks and rivers or around the edge of lakes; they used the edge of roadsides and cross-country ski trails, and they traveled on packed game trails made by a variety of species (e.g., snowshoe hares, red squirrels, elk, and deer). When I tallied my results, nearly two-thirds of the distance that I tracked these red foxes were on snow surfaces that completely supported their weight, even though these kinds of surface formed less than 10 percent of my study area. The foxes wallowed through soft taiga snow only short distances. Furthermore, after a fresh, heavy snowfall the foxes rested and were inactive for about two days. This time-out seemed to allow the snowshoe hares and other animals to reestablish their trails before the red foxes began to move again.

In this same park, I have tracked river otter, who hop and slide along on their belly surfaces, often for hundred of yards. Three hops and then a slide, three hops and then a slide. Their Nordic technique is as old as the northern woods itself. I have observed portions of snow-covered rivers where otters maintain and reuse the same trails over the course of the entire winter, reestablishing their groomed ski trails from dive hole to dive hole after each fresh snowfall.

Prince Albert National Park is one of only two Canadian national parks with a free-ranging bison population (Wood Buffalo National Park is the other; however, reintroduction projects are being planned for other parks). Wood bison historically occupied much of the North American taiga, and at one time, plains bison from the northern portion of the Canadian prairies migrated northward during autumn to spend the winter in the southern boreal forest. When a herd of bison moves through the forest, it leaves behind a path of churned-up snow, which hardens into a well-packed trail. In Prince Albert National Park, wolves, coyotes, red foxes, lynx, and even deer use such trails. Even when the bison trail is covered with ten inches (25 cm) of snow, its surface is still firm, and these animals return to use it for travel. This illustrates the important point that when any wildlife species is eliminated from much of its range, as the bison has been, ecological relationships that we may be unaware of are disrupted.

Caribou: Sculpted by Snow

Caribou, the deer of the taiga and tundra, is beautifully adapted to the harsh winter conditions of the North. Caribou have evolved proportionally larger hooves and dewclaws than other members of the deer family, and their hooves also spread much more easily, to enhance the snowshoe effect. When a caribou puts its hoof down, it makes a four-point landing, while other ungulates normally touch down with just the two sides of their hoof. The four-point hoof of caribou gives them better stability and traction on slippery snow and windblown lakes. Caribou, in fact—partially because of this stability—can outdistance a wolf pack running across the frozen surface of a northern lake. Wolves often try to chase their prey (moose, elk, or deer) out onto lake surfaces, where they may slip and fall, making them vulnerable to the attack by the wolf pack. These ungulates often avoid these frozen surfaces and flee away from them if wolves are nearby. Not so the caribou; they flee toward frozen lakes, as if they know they can outrun the wolves on that wide open, frozen surface.

Caribou have a superbly insulated coat due to several subtle but important adaptations. The individual hairs of the coat taper at both ends and are slightly thicker in the middle. The slight bulge in the middle of the hair gives the coat a greater density at that point, helping to trap warmth in the closed air spaces next to the skin. In

A family of river otters travels across a frozen lake, using their classic technique of several hops and a slide.

addition, each hair has many microscopic air chambers, similar to bubbles, occurring down the length of the hair. These tiny dead air spaces increase the insulation value of the coat to the extent that in calm weather caribou do not need to increase their metabolism to keep warm until the ambient temperature has fallen below −70°F (−66°C).

Despite the efficiency of their coat, energy reserves are critically important, and caribou have developed a number of important feeding strategies. They browse on twigs but also eat lichens (plants composed of algae and fungi growing together in symbiosis), which grow on the ground or on branches of northern conifers. Caribou are able to smell and locate lichens under the snow and also to digest lichens fully, by using a modified fermentation process and by retaining lichens in their rumen for a prolonged period of time. One problem with surviving on lichens is that, while they are fairly high in carbohydrates, they are low in other nutrients. Caribou in the boreal forest cope with this problem in two ways. First, they also search out and eat scarce protein-rich sedges, which stay green and continue to grow under the snow all winter long; and second, they have evolved sophisticated physiological mechanisms for absorbing and recycling minerals and nutrients from their urine and feces.

Even with all these adaptations, caribou are very sensitive to snow conditions, reacting not only to its thickness but also to its hardness. Bill Pruitt and his students at the Taiga Biological Station have studied the behavior of woodland caribou during the winter for a number of years. Not only in Manitoba but also in the Northwest Territories and Finland, Pruitt repeatedly found that both caribou and reindeer seem to respond to critical thresholds. Specifically, if the surface of the snow becomes too hard, or the snow pack is too dense, or the snow is over twenty-four inches (60 cm) deep, these animals begin to emigrate. When one of the thresholds is exceeded, these animals have difficulty digging through the snow to the ground lichen and other plants. The animals then move until they find an area with an acceptable snow cover and stay there until food sources become sparse or until snow conditions prod them to move again.

Snow Cultures
Snow has also affected the human cultures of the taiga.

Snowshoes and Sleds
Snowshoes are an invention of the Indians of the taiga. They were unknown among the Inuit and Eskimo cultures of the tundra, with its hard-packed, windblown snow hard enough to walk on. On the other hand, the Indians of the boreal forest, with its soft, deep snow, have modified snowshoes to match every task and snow condition. To give just several examples, the round bearpaw snowshoe of the maritime Indians works well on snow that frequently forms crusts—from the many freeze-thaw cycles of a maritime winter. The oversized Alaskan and Mackenzie Valley hunting snowshoe, on the other hand, is perfectly adapted to the soft taiga snow of interior Alaska and the Northwest Territories. This fine-meshed snowshoe can be up to eight feet (2.5 m) in length, allowing it to track through soft snow much like a ski. These

Indians also possessed a smaller, lighter, trail snowshoe, which is used to pack down a trail.

The Inuit (the Canadian aboriginal term for Eskimos) developed the *komatik,* a sled that is elevated about twelve inches (30 cm) above the snow and is made of boards lashed onto solid wooden runners. To reduce friction, these runners are brushed with water (sometimes combined with other ingredients) to form a smooth, glazed surface of ice, which glides easily over hard-packed tundra snow. In the boreal forest during the fur-trading era, the traders and the Dene and other First Nations of the boreal forest perfected the wooden toboggan—flat, broad-bottomed, with a high, curled prow and a canvas-sided carriole to shed the soft taiga snow.

The *giens* is used by the Sami, of northern Scandinavia and adjacent parts of Russia. The terrain in much of Lapland is low mountains, with tundra above treeline and below it heavily forested valleys and lowlands. The giens is beautifully designed for a combination of tiaga snow and tundra snow. About ten feet (3 m) long, it is big enough to hold one person and some cargo. It is pulled by a single domesticated reindeer, connected to the giens by a rope, which runs between the reindeer's legs. On the bottom of the giens, in the center, is a keel about four inches (10 cm) wide and one inch (2.5 cm) deep. In soft taiga snow, the curved sides of the giens are in contact with the snow and help the giens to float. On a crust or hard-packed snow, however, the Sami balances the giens so that it rides on its keel.

Sled Dogs

The breed of sled dog also reflect the difference between tundra and taiga snow. The Inuit developed the long-legged and fast husky. The Dene and Woods Cree developed the Indian dog, or trapper dog, bred for power and endurance. It has shorter legs and a deeper chest than a husky and a long, flexible spine so that it can use a lunging gait through the soft taiga snow. Unfortunately, this dog is becoming rare and is in danger of disappearing entirely. Richard Jobb, a Cree from Southend, Saskatchewan, says, "On a packed trail the racing dogs [huskies] will win every time, but in soft snow, those trapper dogs will outlast and beat the racing dogs without any question."

On dogsleds, the Inuit of the eastern Arctic traditionally use a fan-shaped hitch: Each dog is connected by a separate rope to the front of the komatik, and the dogs spread out in front of the sled, each pulling the load by a separate rope. In the taiga the dogs are hooked up one behind the other, with two ropes, one running down on either side of the dogs and connecting them to the toboggan. The first two or three dogs set the course, breaking the trail and packing it down. Behind them, the stronger pullers actually do the hauling.

The *Quin-zhee*

The different northern aboriginal cultures have also developed different snow shelters. The igloo of the Inuit is made out of blocks of windblown tundra and copes well with the Arctic cold. The *quin-zhee* (QUIN- zhee), invented by the Dene, is adapted to the tiaga, making use of the insulating properties of the soft boreal snow. Bill Pruitt suggests that the best way to appreciate a quin-zhee is to build one and sleep in it. You need 1.5 to 2.0 feet (50 to 60 cm) of api—the soft, fluffy snow of the forest floor. Using a snowshoe or large snow scoop, shovel the snow into one central area

Autumn colors of the boreal forest

Beaver lodge

Shoreline of a boreal lake

A fen with floating sphagnum mat

Gray jay

Bull moose

A wolf crosses a snow-covered lake

Young northern pike

Fireweed growing after a crown fire

River otter

Lakes of the boreal forest

Whooping crane breeding grounds

PRINCE ALBERT NATIONAL PARK/PARKS CANADA

PARKS CANADA/MERV SYROTEUK

Pileated woodpecker

PARKS CANADA/ROSS BARCLAY

Stair-step feather moss

Mouse tracks on new ice

Woods Cree woman

Red fox

Pussy willows in the rain

Traditional Sami sled

Candle ice

PARKS CANADA/STU HEARD

Traditional Southern Tutchone winter house

Pressure ridge on a large boreal lake

PARKS CANADA/MIKE JONES

Ground dogwood

Leopard frog

Dog's tongue lichen

Mountain monkshood

Sparrow-eggs lady slipper

Poisonous amanitas mushrooms

Twig-mimic insect and Alpine arnica

Pitcher plants

Forest fire smoke darkening the sky

Northern hawk owl

White pelican

Common loon on its nest

PARKS CANADA/MERV SYROTEUK

PARKS CANADA/BRAD MUIR

Mule deer

Beaver

Cow moose and her calf

Serotinous lodgepole pinecones

Sphagnum moss PARKS CANADA/D. WILKES

Mature Dall's sheep ram

Snowshoe hare

Alder flycatcher feeding young

Three black bear cubs

A potential "torch" tree

Riverbeauty beside fossilized wood

Snowshoes were invented and refined by the Indians of the boreal forest to match their local snow conditions as well as the job that they were undertaking.

until you have a mound four to six feet (1.2 to 2.0 m) high and twelve feet (3.5 m) in diameter. Round off the pile so that it is dome shaped. Before excavating your quin-zhee, let the snow set, or sinter.

Sintering is an interesting physical phenomenon. When soft taiga snow is shoveled or somehow disturbed and then left alone for an hour or two, it loses its soft, fluffy quality and becomes quite hard and rigid. This transformation is caused by mixing together snowflakes of very different temperatures. If the air temperature is −20°F (−30°C) or colder, the snow near the surface is much colder than the snow closer to the ground. The insulating value of soft taiga snow ensures that this is the case. When the snow layers are mixed together, heat flows from the warmer snowflakes to the colder snowflakes. The heat carries along water molecules, which subsequently freeze and establish new molecular bonds that bridge from snowflake to snowflake. These new bonds cause soft taiga snow to set quite hard after an hour or so. As one might expect, the colder the ambient temperature the faster and harder taiga snow sinters. Thus the colder it is the better the conditions are for building a quin-zhee.

Once the snow has sintered, the quin-zhee can be excavated, beginning with the entrance (usually on the side protected from the wind). The surface of the mound will be fairly hard, but the interior snow is soft and easily excavated. Continue carving, pushing the excavated snow out through the door, until there is a room large enough to spread out one or two sleeping bags. Keep the entranceway small; a low,

The *quin-zhee*, the traditional snow shelter of the Dene people, makes elegant use of the soft boreal snow with its high insulating properties.

small doorway will increase the warmth of the room. The walls of the quin-zhee should be six to eight inches thick (15 to 20 cm)—just thick enough to begin to detect light from the outside. At that thickness, taiga snow provides a safe microhabitat against the Arctic cold, whether for a mouse in a pukak, a family of beaver in its lodge, or a Dene hunter inside a quin-zhee. The Arctic makes brothers of us all.

The floor of the quin-zhee should be free of snow. The more snow that is removed, the less that will melt under your sleeping bag, and the more heat that can rise from the ground over the course of a winter's night.

I have camped several times in quin-zhees, with a blanket over the door when it got down to –30°F (–35°C) at night, and yet I was completely comfortable inside. As the interior of a quin-zhee approaches 32°F (0°C), the ceiling starts to drip, and the air becomes warm and humid. A small vent can be carved into the ceiling and the door covering adjusted to keep the inside at approximately 25°F (–5°C), a temperature that will keep the interior snow dry.

Faye Campbell, a Chipewyan from northern Saskatchewan, shared with me a quite different technique for building a quin-zhee. When her father and uncle are out on the trail and stop for the night, they take all the parcels and duffel bags out of their toboggan or snowmobile sled and mound them in a heap. Then they spread a large robe made of beaver pelts sewn together over these bundles, with the fur facing outward. They cover the whole robe with about eight inches (20 cm) of snow and let it set for an hour or so while they enjoy a fire and a meal. Finally, they dig their entranceway into the mound, lift the edge of the beaver robe, and remove the bundles. They are left with a spacious, hide-lined sleeping chamber, snug against the taiga cold. The next morning when they break the quin-zhee down, they shake the snow out of the beaver robe and are on their way.

At Forty Below, It's *Vestis Virem Facit* (Clothes Make the Man)

Winter is an important force in simplifying the taiga's species—reducing plants and animals to those few species that can survive the subarctic cold. Ungulates, if they survive until spring, do so with depleted fat reserves. So taxed is their energy budget that if the growth of new leaves is delayed two weeks by cool spring weather, their winter-related mortality may increase by half. Other species, such as blue jays and raccoons, tend to expand northward until a severe winter eliminates them from these northern areas.

Winter in the snow forest is a fascinating time, both dreadful and wonderful. It can be so still and windless that you suspect that even the air has congealed. I have stood in the boreal forest on frigid nights with a light dusting of new snow and my ears have rung from the absolute silence. At 45° below freezing, the lightest breeze—one that does not disturb even the smoke rising from the chimney—can freeze your face and fingers, turning them a pale, ghastly white in minutes if they are not properly protected.

During December or January the temperature can hover around –40° for weeks on end. These cold spells are trying times. The inside walls of our homes seem to suck heat from our skin at night, and I often sit and ponder how the birds and mammals in the surrounding forest survive these horrible nights.

There is nothing ambivalent about −40°. Either you have the resources to cope with it, or you are soon dead. There is no such thing as "gutting it out"; there is no just barely making it through a frigid taiga night.

Perhaps it is the late winter blizzards that I fear most. I have had the misfortune to be caught out in a few. By spring, the snow is largely transformed into ice crystals, so there is no decent quin-zhee snow around. Sometimes the temperature goes to rock bottom, while whiteouts make travel impossible. I have had to hole up in a tent for three or four days during these storms, tending a fire in the day and simply trying to keep body and soul together in the night. In those long, bitterly cold nights dreams become so vivid that one is tempted to rush headlong into that other, more pleasant reality. I have felt that dark cold—that final, oppressive cold—that settles just before first light. On nights such as these, it is clear that we should take care. They may be right, those who say that the universe is not a friendly place. Our galaxy is for the most part a frigid space, not impressed with the minuscule achievements of life. Even our solar system is mostly barren and cold. We should be careful how we treat this fragile membrane to which we cling. Our biosphere is a freakishly warm microhabitat suspended in a sea of cold.

General References

Aitchison, C. W. 1978. Spiders Active under Snow in Southern Canada. *Symp. Zool. Soc.* (London) no. 42: 139–48.

Calef, G. W. 1981. *Caribou and the Barren Lands*. Ottawa: Canadian Arctic Resources Committee.

Henry, J. D. 1997. *Red Fox: The Catlike Canine*. Rev. ed. Washington, D.C.: Smithsonian Institution Press.

———. 1997. *Foxes: Living on the Edge*. Minocqua, Wis.: NorthWord Publishing.

Kelsall, J. P. 1969. Structural Adaptations of Moose and Deer for Snow. *Journal of Mammalogy* 50:302–10.

Lynch, W. 2001. *The Great Northern Kingdom: Life in the Boreal Forest*. Markham, Ont.: Fitzhenry and Whiteside.

Marchand, P. J. 1987. *Life in the Cold: An Introduction to Winter Ecology*. Hanover, N.H.: University Press of New England.

Penny, C. E., and W. O. Pruitt Jr. 1984. Subnivean Accumulation of CO_2 and Its Effects on Winter Distribution of Small Mammals. *Special Publication* no. 10 (Carnegie Museum of Natural History): 373–80.

Pruitt, W. O., Jr. 1960. Animals in the Snow. *Scientific American* 202:60–68.

———. 1970. Some Ecological Aspects of Snow. In *Ecology of the Subarctic Regions: Proceedings of the 1966 Helsinki Symposium*. Paris. UNESCO.

———. 1984. Snow and Living Things. In *Northern Economy and Resource Management*, edited by R. Olson, J. Geddes, and R. Hastings. Edmonton: University of Alberta Press.

Schnell, R. C., and G. Vali. 1972. Atmospheric Ice Nuclei from Decomposing Vegetation. *Nature* 236:163–65.

Telfer, E. S., and J. P. Kelsall. 1984. Adaptations of Some Large North American Mammals for Survival in Snow. *Ecology* 65:1828–34.

8
A Hare's Breath from Death

As an ecology student, I saw the graphs constructed from the annual fur returns of the Hudson's Bay Company from the early 1800s to the present. The two lines on the chart looked like they were chasing each other in a zigzag race across the page—the one line representing the number of snowshoe hares caught each year and the other line representing the annual tally of Canadian lynx pelts purchased from trappers.

These lines reflected the changes in population levels of the two species. From the peak of abundance to the valley of scarcity, every ten years or so the snowshoe hare population would grow and then decline. Following the hares by a year or so, their main predator, the Canadian lynx, also built up in number and then died off. After that, both populations stayed low for three or four years, and then the cycle began again. It has been going on, as far as we can tell, for all of recorded history, a fundamental rhythm of the North American taiga. This ten-year cycle most dramatically affects the hare and the lynx but also affects other animals, such as the red fox, the great horned and great gray owls, and the ruffed and spruce grouse.

The carefully researched statistics of the cycle speak for themselves: The snowshoe hare population exhibits continentwide cyclic fluctuations. Peak densities occur every eight to eleven years, with an average of ten years; these densities fluctuate approximately five- to twenty-five-fold during a cycle (in some favorable habitats by even more). In a study in northern Alberta, at the peak of the snowshoe hare cycle there were 5,200 hares in a square mile (2.6 km^2) of forest. After the crash, the same patch of forest contained 50 hares.

Peaks in the cycle normally occur at the turn of each decade, for example, 1960–61, 1970–71, 1980–81, and 1990–91, with the lowest densities typically occurring three years after the crash. This is the pattern from central Alaska across the Prairie Provinces to northern Quebec. The degree of synchrony in the cycle across North America is truly impressive.

There are a few places where snowshoe hares are out of synch, Newfoundland being one of them. Newfoundland has its own time zone, so, as one often hears on the CBC, things happen "a half an hour later in Newfoundland," and snowshoe hares seem to observe this tradition. For reasons we do not yet understand, the hares' cycle on the island of Newfoundland appears to peak a year or two after the rest of the continent. The snowshoe hare was introduced into Newfoundland in the 1860s and 1870s. Its population increased rapidly, then crashed, and has shown a ten-year cycle ever since, albeit a year or two behind the rest of the country. The rapidity with which they showed a cycle after being introduced into a new area suggests that, wherever the species occurs in the boreal forest, the characteristic population cycle is likely to appear.

The above are dry textbook statistics. They are impressive enough, but it is quite a different thing to actually experience the die-off of hares. In 1970 I was working in northern Alberta on a black bear management project. In Edmonton, the supervisor of the project had mentioned that he believed the hares were fairly numerous in the study area, perhaps close to their peak. Close to their peak? That was an understatement. The woods were crawling with them. It was autumn; fallen leaves covered the ground, and one listened to the bustle of hares all night. Each red fox and coyote I spotted had a portion of hare dangling from its mouth.

After the bears began their winter sleep I left the area and did not return until the following spring. It did not take long to see what kind of winter it had been. Every aspen branch within hare's reach, every sapling with smooth young bark, had been stripped clean. Young trees and shrubs had been girdled with the chisel marks of hare teeth and were now awaiting a springtime death. Young jack pine seedlings were browsed off wherever they had stuck up through the snow. As the snow melted and pulled back, the true extent of the devastation became clear: Hare carcasses, limbs, and body parts were scattered everywhere; feet and skulls and chewed-off ears littered the forest. And at night, not a shuffling foot, not the scurrying of a single snowshoe hare could be heard. The forest now was as silent and ominous as a death camp. A horrific event had taken place here—once again.

One afternoon that spring, I took a large plastic bag into the forest and in less than half an hour filled it to overflowing with decomposing hare remains. I hiked through many areas of the forest, and they were for the most part the same: The hare population had crashed. For the rest of the summer and for several summers after that, snowshoe hares appeared to be exceedingly rare. It was indeed a wildlife holocaust, one I shall never forget.

The ten-year cycle is not always that macabre. A decade later, in 1980, I was living in northern Saskatchewan. The snowshoe hare population in that area did not reach the abundance or crash as low as in the cycle I witnessed ten years earlier. As one old trapper put it, "The army worms and dung shit saved the day!" Let me explain. During the last two years in that area of Saskatchewan there had been a massive outbreak of tent caterpillars. From mid-May until mid-July, when the caterpillars finally entered their cocoons, tons of droppings fell like black rain from the millions of caterpillars that defoliated every aspen and birch tree within their reach. These droppings had a major fertilizing effect on the herbs, shrubs, and saplings growing on the forest floor, and this lush growth provided a year-round food source, which may have softened the crash of the hares during the winter of 1981. This cycle taught me that each cycle is intricately linked to other events in the forest, that each cycle is a variation on the theme, with its own individual idiosyncrasies.

The Life of the Snowshoe Hare

The snowshoe hare (often called bush rabbit, or simply rabbits, by many Canadians) is one of the most common mammals and one of the dominant herbivores of the North American boreal forest. The range of *Lepus americanus* coincides with the boreal forest, but the species also occupies the Appalachian Mountains as far south as North Carolina and the Rocky Mountains as far south as New Mexico. Their cycle is observed across the taiga and, to a lesser extent, south of there. Human activities in the forests appear to affect the snowshoe hare cycle; the hares in Wisconsin, for example, have not shown a strong cycle since 1950. It is unclear whether the southernmost populations of the snowshoe hare still cycle. Their populations fluctuate, but whether these fluctuations are true cycles is not known. Other than Lloyd Keith's observations from Wisconsin, very little information has been collected south of the U.S.-Canadian border.

A puzzling question for which we do not have an answer is why hare species in the taiga of Scandinavia, Finland, and western Russia do not show population cycles. Both the European hare *(Lepus europaeus)* and the mountain hare *(Lepus timidus)* inhabit these northern forests, but neither species has been reported to show a clear population cycle. On the other hand, *Lepus timidus* shows a ten-year cycle in the Siberian taiga. Why the same species shows a prominent cycle in the Siberian taiga but not in the taiga of Finland and Scandinavia is an intriguing question awaiting future research.

Snowshoe hares are well adapted to the harsh winter conditions of the taiga. Their large, well-furred hind feet allow them to move easily over the soft taiga snow. During winter additional fur grows on their hind feet and between their toes, not only keeping the feet warm but also increasing the snowshoe effect. In soft snow the hare spreads the four long toes of each hind foot widely to help keep itself buoyant in the snow. Foot loading of body weight on large spreading paws has been measured, and only the wolverine exceeds the snowshoe hare in its capacity to stay buoyant in the snow. Another notable snowshoe hare adaptation to the taiga is its ability to change color with the season, turning white during winter and a brown-gray during summer. The snowshoe hare accomplishes this transformation by undergoing three molts annually—in spring, late summer, and midautumn; these molts are triggered by changes in day length. The snowshoe hare's pelage consists of three hair types: guard hairs, pile, and fur. A shift in the color and proportion of these three types of hair changes the animal from dusty brown to pure white and back again at about the same time that its world is changing color.

Rabbits and hares are in the same taxonomic family, Leporidae, but show some differences and belong to different genera. Rabbits' young are born in underground burrows and are naked, blind, and helpless at birth. Hares are born in a surface nest, or form, eyes wide open and fully furred. There are twenty-six recognized species of hare worldwide, with *Lepus americanus* being the smallest in size. Adult hares range in weight from approximately 2.6 to 4.0 pounds (1.2 to 1.8 kg), being largest in Alaska, the Yukon, the eastern United States, and eastern Canada and smallest in

Washington and Oregon. They show their peak weight in November and lose about 14 percent of their body weight over the winter. Adult females tend to be slightly larger and heavier than adult males.

The length of daylight, in addition to being one of the main factors controlling its molt, also influences the reproductive cycle in the snowshoe hare. The rapid increase in snowshoe hare numbers that leads to peak densities is obviously a result of the hare's enormous reproductive potential. Each female over six months of age is capable of producing between eight and eighteen young every summer. To achieve this feat, snowshoe hares begin breeding by the first of April (sooner in the south) and, after a gestation period of thirty-five to thirty-seven days, give birth to their leverets. Females give birth to their litters synchronously, most likely as an evolutionary tactic to "swamp" predators with newly born leverets and thus reduce losses.

Wasting no time, the females breed within hours of giving birth, and the next wave of litters are born five weeks later. Females give birth to as many as four litters per year (fewer during the population decline), with the first litter usually the smallest and the second litter usually the largest. Typically, the newly born leverets hide together under a fallen tree trunk or in a tangle of branches or shrubs for the first three to five days. Then they hide separately. The females are with their young for only about ten minutes each day, during which she gives them their once-a-day nursing. The female stops nursing her young after a few weeks. Because hares are

The snowshoe hare is a keystone species in the North American boreal forest because its ten-year cycle affects so many other animals and plants.

not territorial, these weaned juvenile hares begin to forage on the same forested terrain as the adults.

The distribution of snowshoe hares in the forest depends very much on the amount of dense cover provided by young trees or shrubs. When in an experiment researchers removed the shrub layer from an aspen stand, the hare density reduced by more than 75 percent. As their populations begin to build, snowshoe hares are forced to occupy more mature forests, even those lacking a dense understory. As their population peaks, they are distributed widely and are common even in open areas that offer abundant food. As the population declines, they occupy areas of dense cover, perhaps in an effort to avoid the still abundant predators. As the population declines further, they inhabit patches of heavy willow or dense spruce, and it is these pockets of hares that form the nuclei for the next cycle.

In summer snowshoe hares eat forbs, grasses, leaves of shrubs and trees, as well as some woody browse. In winter they eat twigs and some bark of shrubs and trees and also dig through shallow snow to feed on forbs and grasses. The unique way hares process their food is called coprophagy (eating excrement). It allows snowshoe hares to survive on low-protein browse by eating a lot of it and passing it through their digestive system twice. They excrete two types of droppings. The hard, brown droppings are their waste products. The soft, green droppings are food balls infected with bacteria. These food balls have spent several hours in the hare's caecum (a pouch at the beginning of the large intestine), where the bacteria has been partially digesting resistant plant material. When a hare excretes one of these green pellets, it is eaten again immediately so that it can pass through the hare's digestive system a second time, where most of the altered plant material and some of the protein-rich bacteria are digested and absorbed. One ramification of this feeding strategy is that hares eat the same amount of food daily—they can process only so much plant material through their digestive tract each day. Consequently, when the quality of the available plant forage drops, hares cannot compensate by eating more. Interestingly, even in winter, hares maintain only small fat reserves, which can sustain them for only four to six days without eating. They appear to eat approximately 0.75 pounds (300 g) of browse each day, and they are better able to maintain their body weight on smaller twigs than on larger twigs, which have more wood and less bark.

A number of predators prey on snowshoe hares. The main predators of adult hares are Canadian lynx, coyotes, red foxes, goshawks, and great horned owls. In contrast, young hares are predominantly preyed upon by small raptors (boreal owls, red-tailed hawks, kestrels, northern hawk owls) and even by such unexpected predators as red squirrels, Arctic ground squirrels, and weasels.

Because the hare is the dominant herbivore of the taiga and is preyed upon by so many species, some researchers identify it as the keystone herbivore of the boreal forest. That is, it is like an archway keystone, which holds the other blocks in place: A marked change in its population causes significant changes in the populations of many other species in the ecosystem. When its population crashes, the populations of several predators, such as Canadian lynx, coyote, red fox, and great horned owl, are known to decline as well. Reproduction in these predators is reduced to a fraction of what it is at the peak of the snowshoe hare cycle, so that a year or two after the hare population declines, many of these predator populations are at low density as well.

SMITHSONIAN NATURAL HISTORY SERIES

The Canadian lynx is one of the main predators of snowshoe hares, and when hares are scarce it will travel long distances searching for this prey. These long-distance movements by lynx help to synchronize the hare-lynx population cycles over large areas.

Documenting—and Explaining—the Cycle

Trappers have known about the snowshoe hare cycle for decades, and native peoples have known about it for centuries. A person cannot live in the boreal forest and be unaware of the dramatic changes in the lives of snowshoe hares and their predators. In one part of the cycle, whenever you are out in the forest you encounter the packed runways of snowshoe hares and one or two sets of lynx tracks. Several years later you may go three or four winters without seeing a single lynx track, and seeing the Y-shaped tracks of one snowshoe hare can be noteworthy. These changes are fundamental rhythms of boreal winters.

Scientists first documented the snowshoe hare cycle in the early 1930s. Beginning in 1931 Charles Elton, later joined by M. Nicholson and the Canadians Dennis and Helen Chitty, all of Oxford University, mailed questionnaires to Hudson's Bay Company posts, Royal Canadian Mounted Police posts, and game officers in the different regions, asking whether hare numbers were increasing or decreasing relative to the previous year. The responses of approximately 500 observers were compiled into maps for each year of relative abundance of snowshoe hares, from 1931 to 1948. Dale MacLulich built on these early efforts, using the records kept by factors of various Hudson's Bay Company trading posts. He documented the number of snowshoe hare pelts collected at each post as well as the number of lynx pelts purchased each year from 1845 to 1935. He produced the aforementioned graph (which appears in almost every ecology textbook) of the buildup and crash of the snowshoe hare population every decade, with the lynx population following a year or two behind. This graph traces the very heartbeat of the boreal forest.

A ten-year population cycle must be one of the most difficult ecological phenomena to study. It takes a decade to get an initial look at the full cycle. If you missed something or wish to have another look at some aspect, you have to add another ten years onto your research project. In addition, each cycle shows its own idiosyncrasies, so that a researcher never examines the same cycle twice. And what exactly do scientists want to understand about this ten-year drama between the hares and their predators? The challenging questions are: What causes the sudden decrease of the snowshoe hare population and the subsequent decrease in the numbers of their predators? What keeps hare numbers low for three or four years? And last, why doesn't the whole system reach stability—that is, why don't the hare population and the predator populations oscillate until they reach equilibrium?

Let's consider that last question first. The persistence of the cycle over centuries is believed to have a lot to do with the simplicity of northern ecosystems. The snowshoe hare and the lynx come pretty close to being a one-prey, one-predator ecological system. Because there are few buffer prey species to sustain the predators once the hares are depleted, the predator population, after a year or so, decreases along with its prey.

In laboratories under controlled conditions, when one-prey, one-predator systems are built and maintained (for example, with house flies as prey and wasps as predators), these two populations often develop dramatic cycles, with the predator lagging shortly behind the prey. Such laboratory studies offer evidence that cycles are characteristic of simple ecosystems, the kind of ecosystem found in the North. Two widespread northern cycles are the ten-year snowshoe hare–lynx cycle of the boreal forest and the four-year lemming cycle of the Arctic tundra.

Robert May—building on the work of Patrick Moran and John Maynard Smith—provides an important theoretical insight into the dynamics of the snowshoe hare cycle. He notes that the cycle is maintained by the snowshoe hare population staying low for three to four years after its decline. If the hare population recovered more quickly, then both predator and prey populations would tend toward equilibrium, and the cycle would disappear. May's insight is built on the theory that a time lag helps perpetuate a population cycle. Why is this? He argues that there are four inherent stages of a predator-prey cycle: low prey, low predator; high prey, low predator; high prey, high predator; and low prey, high predator. If the time lag is approximately one quarter the length of the whole cycle, as it is in the snowshoe hare–lynx cycle (approximately three of ten years), then the time lag helps move predator and prey to the next phase of the cycle rather than toward equilibrium. Populations that respond quickly to changes in their environment generate stability; populations that respond only after a time lag tend to generate cycles.

Lloyd Keith has been intensely researching the snowshoe hare cycle since the early 1960s, which means that he has observed four complete snowshoe hare cycles. His understanding of its causes has developed and changed over the course of his forty years of research. During the 1970s Keith believed that an interaction of food supply and predation causes the decline of the snowshoe hare population. Furthermore, he believed and still argues that changes in winter weather can improve or worsen the hares' die-off, and thus winter weather helps synchronize the cycle over large geographic regions.

More recently, field studies in Alaska, the Yukon, and on Keith's study area in northern Alberta demonstrate that 80–90 percent of hare deaths at the peak and during the decline of the cycle are caused by predation, mostly by Canadian lynx, coyotes, and great horned owls. These results led Charles Krebs, Tony Sinclair, Stan Boutin, and their co-workers to suggest that predators drove the snowshoe hare cycle from its peak density into population decline. For example, this group found that, on their southwestern Yukon study area, during the decline phase of the cycle less than 1 percent of the hares survive an entire year, and predators are known to be responsible for at least 83 percent of these deaths. Results such as these have led Keith to agree with Krebs and colleagues and to conclude that predation alone is sufficient to reduce hare numbers during the decline phase, even where local food resources for the hares remain relatively high.

Although Keith believes that predators drive the snowshoe hares into population decline, he still argues that vegetation plays an important role. Food supplies at the peak of the cycle, he believes, become somewhat reduced on a regional basis, forcing hares to feed in areas with less shrub cover, which makes them more vulnerable to predation. As hares are hunted out at these sites, predators switch to other locations, so that predation pressure becomes high across the region. Great horned owls and Canadian lynx, he says, will move long distances to wherever hares remain abundant, and this factor together with harsh or mild winter weather synchronizes the snowshoe hare cycle across the continent.

To these changes we can also add changes in the reproductive biology of the snowshoe hare. Rudy Boonstra, a professor at the University of Toronto, has been part of the Kluane Lake research team in the Yukon for the past twenty-five years. In a recent interview with me, he explained the interactions between predators, snowshoe hares, and changes in the hare's reproductive biology:

> Predators such as coyotes, lynx, and great horned owls keep building up in numbers during the peak phase of the cycle and are continually preying on hares. The predators finally drive the hare numbers down. However, there is another important process going on that helps to drive the hare numbers down even further. During the peak phase of the cycle, snowshoe hares become stressed. Mark O'Donaghue from our research team has documented that when coyotes and lynxes hunt snowshoe hare, they are successful only about 30 to 40 percent of the time. These failed hunting attempts stress the hares. Furthermore, through seeing and smelling the tracks and scent marks of predators, the hares are aware of the greater number of predators in their area. Stress alters the level of various hormones circulating in the blood of the hares, and these hormones alter the reproductive physiology of the hares, possibly on a permanent basis. Even before the peak of the cycle, changes in the reproductive output of female hares begin to appear. The fourth litter and then the third litter that females have earlier in the cycle disappear, and the number of leverets per litter declines. We see these reproductive changes before and during the peak density, but these changes intensify throughout the decline of the

snowshoe hare population. It is a combination of lower reproduction and heavy predation that drives the snowshoe hare down to its lowest level.

In addition, Boonstra suggests that these hormonal changes may be slow to disappear. He speculates that predator-induced stress ultimately operates at the level of the brain and that these detrimental effects on the brain affect reproductive function, possibly through placental transfer of hormones to the developing fetuses. Consequently, he believes that the critical three- or four-year time lag, when the snowshoe hare population stays low, is caused by the fact that it takes this long to establish in the population females that do not have stress-altered physiologies and thus show greater reproductive output. It is only after this time period that snowshoe hare numbers begin to dramatically increase.

Sunspots and Synchrony

During the early 1930s several researchers suggested that snowshoe hare cycles were caused by cycles of sunspots, huge dark circular areas that appear on the surface of the Sun approximately once each decade—and then disappear. Sunspots cause changes in the climate of the boreal forest, they said, which then affects the snowshoe hare cycle. In the late 1930s Dale MacLulich argued against the theory, and Moran and Keith gave evidence that, while the hares cycle on a ten-year average, the sunspots cycle on an eleven-year average. Over the course of a century the two cycles would be badly out of synch.

More recent research by astronomers documents an average of 10.6 years for the sunspot cycle—fairly close to the snowshoe hares. A. R. E. Sinclair and J. M. Gosline, of the University of British Columbia—together with other researchers from the Kluane Lake research station—show that severe browsing of white spruce by hares leaves dark marks in the annual growth rings of these trees. By analyzing these dark marks, they were able to document twenty-three cycles from 1751 to 1983. Using modern correlation techniques, they show that peaks of the snowshoe hare cycles are indeed correlated with peaks of sunspot activity. Furthermore, the cycles of hares and sunspots were in phase during three periods—1751 to 1787, 1838 to 1870, and 1948 to 1983. Sinclair and Gosline suggest that sunspots synchronize the snowshoe hare cycle across most of North America during the thirty-to-forty-year period that the sunspot cycle is at its maximum activity, thus increasing the solar irradiance to the Earth, which affects global weather patterns. Snow records from ice cores taken on nearby Mount Logan, Canada's highest mountain, demonstrate that a peak in snow accumulation correlates with sunspot activity. During the fifty to seventy years of low sunspot activity, snowshoe hare cycles drift out of phase with sunspot cycles, but they are brought back into phase at the next period of high sunspot activity—an event that occurs every eighty to ninety years.

Sinclair and Gosline stress that sunspot activity does not cause the hare cycle but that it provides a phase-locking mechanism synchronizing the hare cycle across the continent. The actual mechanisms by which sunspots affect climate—which in turn synchronize snowshoe hare cycles across North America—have yet to be discovered. Future research on these mechanisms should produce interesting results.

The Taiga's Chemical Warfare

An alternative view of what causes the snowshoe hare cycle is provided by John Bryant, a biologist from the University of Alaska. Working with David Klein and others, Bryant concludes that the ten-year cycle is caused by a form of chemical warfare orchestrated by the trees and shrubs that snowshoe hares browse during winter.

To begin to understand this interesting theory, we start by looking at green alder, a common shrub found throughout the North American taiga. *Alnus crispa* is a strange plant in that the microorganisms that live on its roots have an alchemistlike power of taking nitrogen gas straight from the air and eventually releasing nitrates into the nitrogen-poor taiga soils. Not only does green alder play an important role in enriching taiga soils but it also offers some of the most protein-rich browse available during food-scarce boreal winters. The surprise is that green alder is virtually never browsed. Why? What keeps it from being nibbled upon by moose, grouse, caribou, or mice? Bryant's research shows that the twigs and buds of green alder carry relatively large amounts of several toxic substances. One of the toxic metabolites found in the bark and buds of green alder is pinosylvan methyl ether, a chemical known to act as a strong repellant to a wide range of herbivores. In fact, in the tropics it has been shown that if this chemical is extracted from certain local plants and used as a wood preservative at extremely low concentrations, termites do not touch the wood for at least two years. Bryant suggests that in the taiga the same toxic substance keeps green alder from becoming winter browse.

It would appear that the boreal forest is not such an edible world as it might seem. Analyses of twigs from other tree and shrub species show that other repellant chemicals are frequently present. Those defensive chemicals involve terpene and phenolic resins, which Bryant believes keep these other plant species from being browsed by snowshoe hares and other taiga herbivores.

The theory that Bryant and his colleagues put forward is complex, but it can be summarized as follows: At the peak of the snowshoe hare cycle, certain woody plants respond to severe browsing by producing toxic metabolites in their twigs and buds, a response that may play a significant role in the regulation of the snowshoe hare population. To understand the theory, we divide the taiga into two groups of plants: late successional plants, such as black spruce and green alder, which reproduce under an established canopy; and early successional plants, such as aspen, paper birch, and willow, which usually reproduce in the open.

Bryant claims that late successional plants maintain their chemical defenses almost continuously and that there is a good reason for this strategy. Plants that grow in habitats such as black spruce moss yards or muskeg communities must be adapted to low-nutrient environments, that is, areas where decomposition is suppressed by acidification and low soil temperature. Consequently, these late plants, because of poor nutrient conditions, take a number of years to grow beyond the reach of most browsing mammals. To protect themselves, they maintain a high level of toxic metabolites in their woody parts. This requires a major investment of energy, but it is necessary—once their twigs, bark, or buds have been browsed, they are not easily replaced. Furthermore, because the soils are so poor, late plants often use carbon-based metabolic by-products as their chemical defenses instead of one requiring nitrogen. In essence, late plants specialize in highly repellant carbon-based resins to protect themselves against herbivores.

Early successional plants' version of chemical warfare is more complex. They usually sprout or seed-in immediately after a major disturbance. For example, a forest fire opens the canopy, bringing abundant sunlight. Furthermore, a fire releases a lot of nutrients into the soil from the ash, so the site is rich with the nutrients that early plants need to support their rapid growth. Unlike late successional plants, early successional plants grow in resource-rich environments. Within a couple of years, they often grow beyond the reach of most herbivores.

In a recently burned area, aspen, jack pine, and birch compete to establish themselves as the dominant tree in the postfire canopy. Consequently, Bryant believes that early successional plants are genetically programmed to allocate their resources between two priorities: first, vertical growth, to outcompete other trees and establish themselves as a part of the new canopy; and second, maintenance of chemical repellants in their young stems, twigs, and buds, to protect themselves against the browsing of snowshoe hares and other herbivores. When they grow beyond the reach of browsers, they are programmed to redistribute incoming nutrients and energy resources into the lateral development of their crowns and into the production of seeds, the latter helping to ensure the survival of their genes into future generations.

Bryant points out that woody plants from many regions have in fact two growth forms: a juvenile strategy and a mature strategy. Many juvenile woody plants produce antiherbivore devices, such as thorns or toxic chemicals, to prevent their young stems from being browsed. These devices often disappear when the plants mature. Severe pruning often causes a mature plant to revert to its juvenile form—for example, sending out sprouts that carry a high concentration of chemical repellants. Severe browsing of taiga plants by snowshoe hares near or at peak density causes early successional plants to put forth adventitious shoots heavily laden with toxic metabolites. Because hares do not eat these twigs, they starve. Thus, according to Bryant, the chemical warfare of early successional plants triggers the massive die-off of snowshoe hares.

The evidence by which Bryant supports his theory is fairly convincing. For example, he shows that the twigs of juvenile paper birch are covered with tiny resin glands, which are not present on twigs of mature paper birch. Bryant dissolved and extracted resin compounds from these glands and painted highly preferred browse cuttings and food pellets with the solution. He found that captive snowshoe hares consistently avoided the treated food sources even though they were on a severe starvation diet. Bryant and Klein then conducted feeding experiments among wild hares, offering them both mature and juvenile twigs from various tree and shrub species. Consistently, the snowshoe hares avoided the twigs that carried repellant chemicals, be they pinosylvan methyl ether, terpenes, or phenolic resins.

In summary, Bryant's theory asserts that severe browsing by snowshoe hares makes early successional plants revert to juvenile growth forms, which means that the sprouts carry chemical defenses for about three years, after which time the sprouts produce chemical-free lateral branches. Because of this food shortage (orchestrated by the plants themselves), the hare population is forced to stay at low densities for three or so years, when the new growth begins to supply snowshoe hares with abundant food. The hares then begin having larger and more frequent litters, and the population builds. Late successional plants, particularly forests of black spruce and river alder swamps, offer a refuge where the small snowshoe hare population can survive. Because these plants maintain their chemical defenses continuously, the hare population stays at a low but steady density in these isolated pockets. In these black spruce forests and alder swamps, a few hare families survive the die-off, and the area is eventually repopulated with snowshoe hares.

Teasing Apart the Causes of the Cycle

Bryant's theory is controversial, and some people have their doubts about it. Although Keith acknowledges that the reduced palatability of juvenile plant growth is well established and that this juvenile growth could cause a three- to four-year time lag in the recovery of hare numbers, he says, "It remains to be shown whether plant chemical defense actually plays a significant role in the dynamics of natural populations. Bryant and colleagues have not worked with hares at the population level." Keith is definitely of the "wait and see" attitude toward John Bryant's theory.

Other researchers—especially Charles Krebs, Tony Sinclair, and Jamie Smith of the University of British Columbia and Stan Boutin of the University of Alberta—question the geographic extent over which Bryant's experimental results apply. Krebs points out that the snowshoe hare cycle is fairly synchronous across North America but that the plants that make up the hare's diet vary greatly from region to region. He believes it unlikely, therefore, that plants drive the synchronous decline of the snowshoe hare. However, the snowshoe hare's predators (lynx, coyote, red fox, goshawk) are nearly the same across the continent, making them more likely candidates for driving the cycle.

In their Kluane Lake study area, these researchers found that snowshoe hares browse on balsam poplar, white spruce, and two species of willow in direct proportion to the twig's resin content, thus completely contradicting what Bryant found in the Fairbanks area of Alaska. Furthermore, from 1976 to 1984 this research team manipulated food supplies for snowshoe hares in their study area, and they demonstrated that the cyclic decline of hares could not be prevented by either artificial or natural food additions. Setting out extra food delayed but did not prevent the die-off of the snowshoe hare. During the peak and decline phases of the cycle, 83 percent of the deaths were due to predation, and only 9 percent were attributed to starvation. These results agree with Keith's findings—that hares rarely starve to death. Krebs and his colleagues also agree with Keith that a food shortage may affect mortality indirectly by driving hares into areas where they are vulnerable to predation.

From 1986 to 1996 these researchers worked with Rudy Boonstra, Mark Dale, Kathy Martin, and Roy Turkington. This team of highly respected ecologists felt that carefully designed field experiments were needed to untangle the causes of the decline portion of the snowshoe hare cycle. They also wished to understand how changes in the cycle affect other animals in the taiga community. In the open, undisturbed boreal forest surrounding Kluane Lake, they set up 0.4-square-mile (1 km^2) experimental plots and randomly assigned treatments to these plots. Three plots were control areas and were not manipulated, two plots were provided year round with nearly unlimited food (commercial rabbit chow); one plot was surrounded by an electric fence that was permeable to hares but that excluded mammalian predators larger than weasels. One plot was surrounded by an electric fence to exclude predators, but food (again, commercial rabbit chow) was added continuously in order to study the interaction between these two treatments, namely, *excluding predators* and *adding food*. Finally, on two plots fertilizer containing nitrogen, potassium, and phosphorus *(adding nutrients)* was added to the soil to stimulate plant growth. The team was not able to exclude avian predators despite netting and monofiliment fishing line strung across the top of the plots.

Problems presented themselves from the beginning. To keep lynx from moving through the electric fencing once snow was on the ground, they set up a barrier of chicken wire inside the electric fence, which acted as a ground. To keep coyotes from digging under the chicken wire, they added a strand of electric wire at ground level (and cut holes in the chicken wire to allow hares to pass through). All plot treatments were in place and effective by the spring of 1989 and were kept active until the spring of 1996. Snowshoe hare density was measured by live trapping and by using radio telemetry to measure survival rates and reproductive characteristics. For five years, through the peak and decline of the snowshoe hare cycle, snowshoe hares on their experimental plots and on control plots were compared.

The *excluding predators* plots doubled—and the *adding food* plots tripled—hare density during the peak and decline of the snowshoe hare cycle. *Excluding predators and adding food* plots maintained a density of hares elevenfold above the control plots. *Adding nutrients* plots showed increased plant growth but no significant effect on hare density or survival. These results are averages for the full period, 1989 to 1996. Even greater differences were observed when the hares were close to the bottom of their decline, when hare densities on the *excluding predators and adding food* plot exceeded the control plots by thirty-six-fold. On the other hand, the *adding nutrients* to the forest ecosystem had virtually no effect on snowshoe hare numbers despite the increased growth of herbs, grasses, shrubs, and trees—the fertilized plots contained virtually the same number of hares as the control plots.

These field experiments support the view that population increase during the peak phase is halted by both increased mortality and reduced reproductive output. The mortality of juvenile hares increases before the peak phase is reached, whereas adult mortality does not become severe until the decline phase. In addition, the decline phase is characterized by poor survival of both juveniles and adults as well as lowered reproduction in females as a result of litter sizes reducing and the third and fourth litters disappearing altogether. The researchers conclude that the snowshoe hare cycle is caused by the simultaneous and interacting effects of food supply, predation, and reproductive output of female hares. Concerning the low phase of the cycle, they state, "Our studies have provided little data on the causes of the low phase of hare cycle which can persist for three or four years. . . . Whether the direct or indirect effects of predation can also explain the low phase remains an open question." They conclude that:

- Predators drive the hare cycle directly by killing and indirectly by causing stress.
- Starvation rarely occurs, and there is rarely a shortage of food for hares in the Kluane Lake study area. However, food shortages may make hares more susceptible to predation.

Arctic ground squirrels have been found to be effective predators of snowshoe hare leverets under two weeks of age.

- Red squirrels are unaffected by the hare cycle, and their population fluctuates instead in response to irregular crops of spruce cones.
- Arctic ground squirrels are strongly affected by hare predators, and their population fluctuates in unison with that of snowshoe hares.
- Red squirrels and Arctic ground squirrels are effective predators of young hares under the age of two weeks.
- Lynx can disperse widely when the hare population declines. This could synchronize the cycles geographically.
- Improved plant production does not translate into increased hare density.

Do These Research Findings Contradict or Complement Each Other?

Research on the snowshoe hare cycle has been marked by Herculean efforts by teams of scientists embarking on projects spanning decades. The research teams led by Keith, Krebs, and Bryant have shown great commitment to uncovering the causes of one of the most complex phenomena found in the boreal forest of North America.

Lloyd Keith and Charles Krebs agree that the peak and decline of snowshoe hares every ten years is caused by an interaction of food supply and predators. The balance between these two factors may vary somewhat from region to region and even from cycle to cycle. Their theory offers the soundest explanation for the peak and decline of the snowshoe hare cycle. Furthermore, some ecologists believe that the Keith-Krebs theory is not in conflict with the Bryant theory—that both theories offer valid explanations for the snowshoe hare cycle. The Keith-Krebs predator-plus-food theory explains the causes driving the decline of the snowshoe hare from its peak to its lowest density.

John Bryant's plant-chemical-defense theory explains how the three- to four-year time lag at the bottom of the cycle is maintained and how the slow recovery of the snowshoe hare population takes place. If Robert May is correct—that this three-year time lag is essential for maintaining the ten-year cycle—then Bryant's theory may explain how vegetation itself determines this part of the cycle: Only after three years does new plant growth appear free of toxic chemicals and only after three years do reproductive changes in female hares (as described by Rudy Boonstra) take place.

Things are far from settled. Boonstra, for example, questions the theory that toxic substances in plants are responsible for maintaining low snowshoe hare populations for three or four years, arguing that the toxic chemicals largely disappear from browsed plants early in the low phase of the hare cycle. Instead, he suggests that long-term effects of chronic stress, affecting adult hares directly or passed on during pregnancy from mother to offspring, are the best explanation for the extended low phase of the cycle.

Yes, things are far from settled, and future research will surely bring new discoveries and new debates. However, in some instances, rather than being contradictory, the theories put forward by Keith, Bryant, Krebs, and their colleagues complement each other. An interaction between food supply and predators causes the snowshoe hare populations to peak and decline, and an interaction between toxic chemicals in plants and the effects of long-term stress on the hares keeps the hare population low for three to four years. At that point the taiga's cycle begins again, affecting snowshoe hares, ground squirrels, Canadian lynx, coyotes, red foxes, goshawks, great horned owls, and other forest species. It is a complex and challenging cycle, but it is truly the syncopated heartbeat of the North American boreal forest. Thanks to these researchers, we are beginning to know how to take the pulse of the taiga.

General References

Boonstra, R., D. Hik, G. R. Singleton, and A. Tinnikov. 1998. The Impact of Predator-Induced Stress on the Snowshoe Hare Cycle. *Ecological Monographs* 79:371–94.

Boonstra, R., C. J. Krebs., and N. C. Stenseth. 1998. Population Cycles in Small Mammals: The Problem of Explaining the Low Phase. *Ecology* 79:1479–88.

Boonstra, R., R. K. Swihart, P. B. Reichart, and L. Newton. 1994. Biogeography of Woody Plant Chemical Defense against Snowshoe Hare Browsing: Comparisons of Alaska and Eastern North America. *Oikos* 70:385–95.

Bryant, J. P. 1981. Hare Trigger. *Natural History* 90:46–53.

Keith, L. B. 1990. Dynamics of Snowshoe Hare Populations. In *Current Mammalogy*, edited by H. H. Genoways. New York: Plenum Press.

Krebs, C. J., S. Boutin, R. Boonstra, A. R. E. Sinclair, J. N. M. Smith, M. R. T. Dale, K. Martin, and R. Turkington. 1995. Impact of Food and Predation on the Snowshoe Hare Cycle. *Science* (New York) 269:1112–15.

Krebs, C. J., S. Boutin, and R. Boonstra, eds. 2001. *Ecosystem Dynamics of the Boreal Forest: The Kluane Project*. Oxford: Oxford University Press.

May, R. M. 1974. *Stability and Complexity in Model Ecosystems*. 2d ed. Princeton, N.J.: Princeton University Press.

Sinclair, A. R. E., and J. M. Gosline. 1997. Solar Activity and Mammal Cycles in the Northern Hemisphere. *American Naturalist* 149:776–84.

Sinclair, A. R. E., J. M. Gosline, G. Holdsworth, C. J. Krebs, S. Boutin, J. N. M. Smith, R. Boonstra, and M. Dale. 1993. Can the Solar Cycle and Climate Synchronize the Snowshoe Hare Cycle in Canada? Evidence from Tree Rings and Ice Cores. *American Naturalist* 141:173–98.

9
Muskegs—Halfway between Land and Water

Sphagnum moss is a primitive plant that violates all the rules. Fossils tell us that the sphagnum genus is at least 90 million years old and has apparently changed little over its evolutionary history. Yet this three-inch-tall tyrant can topple trees and determine what other plants will be tolerated in its domain. By acidifying its environment, sphagnum rules firmly, all the while expending minimum energy on its version of chemical warfare. Most impressive, sphagnum creates its own terrain by slowly accumulating huge mounds of peat moss. The actions of ice crystals slowly sculpt these mounds into muskeg, a *Star Trek* terrain, the most bizarre landform of the boreal forest.

Sphagnum is the master of muskegs. However, at least one environmental requirement must be present before sphagnum can stage a successful coup d'état; namely, precipitation must exceed evaporation. That is, in the region as a whole, the sum total of snow, rain, mist, and fog must exceed the sum total of water evaporated off the land and water transpired through plants. Put simply, total water gained must be greater than total water lost. If this condition is present (as it is over much of the boreal forest), then sphagnum may establish itself and begin to rule its muskeg community.

Why must that prerequisite be met? An excess of precipitation tends to fill large and small depressions with water, and these soggy low spots are prime habitat for the invasion of sphagnum mosses. Sphagnum's takeover happens gradually. Sphagnum, after all, is a primitive moss spreading into new areas mostly by spores. Its initial establishment in one of these water-filled depressions may take decades. Once it does gain a foothold, many more years may pass before it has grown into a large mass of floating moss. However, once this "floating blob" stage is reached, sphagnum starts its water-diversion tactics, blocking the flow of water both into and out of the site. Sphagnum gains quite different advantages from each kind of blockage.

First, as water draining away from the water-filled depression slows and eventually stops completely as the moss grows larger, part of the floating mat may drift downstream, growing thick in outlets. What ecological advantage does sphagnum gain from this? The blocked outlets allow the sphagnum to flood adjacent habitat, thus increasing the size of the water-soaked area and, consequently, the area available for its growth. Second, as water is blocked from coming into the site by the growing mat, the moss grows slowly upstream, until the small channels and rivulets that bring water into the site are also blocked. What is the ecological advantage to sphagnum from this diversion? The surface water diverted to other areas carries with it calcium, phosphorus, nitrogen, and other minerals necessary to the growth of competing plants. With the supply of minerals effectively cut off from shrubs and many vascular plants, sphagnum begins to dominate the site. Nor does sphagnum stop there. As it continues to accumulate on the site, a layer of peat moss covers the mineral soil on the bottom of the depression, further sealing the site off from its normal supply of minerals.

A wetland in this plugged-and-sealed condition has been turned into an ombrotrophic wetland, meaning *fed only by rain*. The only minerals it receives are dust and clay particles blown onto it by wind or deposited by rain. In ombrotrophic wetlands, minerals are not carried onto the site through the normal movement of surface water, as they would be in minerotrophic wetlands, that is, those with a normal, unblocked, flow-through drainage. As a result of this more generous supply of minerals, a minerotrophic wetland is more productive than a sphagnum bog.

Bogs and Fens and Soaking Feet

Whether a marshy site is ombrotrophic or minerotrophic usually determines which of the two fundamental types of wetland will develop. These two kinds of wetland are known as bogs and fens. As a hiker in the taiga, it's easy to know whether you are standing in a bog or in a fen—measure the amount of water in your rubber boots!

Bogs offer fairly firm footing because most of the accumulated growth of sphagnum moss does not rot away. In a bog, decomposition is virtually at a standstill, so the dead vegetation simply piles up year after year. Fens, on the other hand, offer treacherous footing. The good supply of minerals as well as the oxygen-rich water in fens permit constant digestion of the dead vegetation by bacteria, water insects, certain beetles, and fungi; the bottom is continuously rotting out from under the floating shelf of living vegetation. Fens are those legendary quaking mats of vegetation that are reputed to periodically swallow bulldozers or boxcars in one protracted gulp. Hiking through a fen is like hiking through a flower-laden meadow perforated with open manholes. An experienced fen walker knows that it is only a matter of time before he or she drops out of sight to come up festooned with the fetid false bottom that forms underneath the quaking mat. However, this baptism by organic debris is never as bad as one remembers it to be, and while I can't exactly recommend the experience, I must say it is a good way to cool off and gain some relief from the hordes of mosquitoes and black flies that inhabit muskegs.

Bogs operate quite differently. Sphagnum is not a single plant species but a group of fifty closely related moss species. Each favors slightly different moisture and growth conditions. Some sphagnum mosses are best at establishing themselves in water-saturated pools, and others grow best on the relatively dry hummocks that develop into purple or chartreuse cushions in a mature muskeg. However, all mosses

A dragonfly hatching from its larval skin will soon take flight. Muskegs and fens are highly productive habitats for boreal insects.

of the sphagnum genus have certain characteristics in common. Once the drainage is blocked and the supply of mineral- and oxygen-rich water is cut off, they begin to produce acid, killing off all plants that cannot tolerate such conditions. The lack of oxygen from the water-saturated soils and the buildup of acid also ultimately kill many of the decomposers: the bacteria, scavenging insects, and fungi. Decomposition is brought close to a standstill, and dead plant material as well as other organic debris begins to accumulate.

Many fen-loving plant species such as sedges (certain *Carex* species) and buckbean *(Menyanthes trifoliata)* cannot survive in bogs. Labrador tea, leatherleaf, certain orchids, and black spruce are among the few acid-tolerating plants that can grow there. However, bog plants do grow, but at approximately one-sixth the growth rate of an aspen forest. Even at this reduced rate, live and dead plant material starts to accumulate, and because little of this plant matter decomposes, mounds of organic material soon begin to collect. As the peat accumulates, the bog takes on a raised, or domed, profile. The center of a bog can be elevated up to fifteen feet (4.5 m) higher than the surrounding forest floor. Like a huge sponge, this raised bog is often saturated with water, which is held in place by the millions of microscopic spaces in the sphagnum peat. The raised profile of such a bog further isolates plants growing on its surface from an adequate supply of soil minerals. In essence, sphagnum has evolved so that masses of sphagnum break the food chain that normally, in more productive ecosystems, recycles nutrients. By acidifying and breaking these interrelationships, sphagnum causes organic material (mostly dead sphagnum moss) to amass. A site dominated by sphagnum becomes organic terrain—derived from living plants and accumulated dead plant material rather than from mineral soil and bedrock. As deposits of peat build, a number of organic landforms emerge. These forms become more exotic the farther north one goes.

Palsas and Peat Plateaus

Bogs and fens are adequate terms if you don't venture into northern Canada, Alaska, or the northern reaches of Scandinavia or Siberia. But venture farther north, where the imbalance between precipitation and evaporation is even more exaggerated and permafrost enters the picture, and you will discover new landforms not encountered along the southern border of the taiga.

Consider, for example, the huge, deep peat slopes that Joseph and James Tyrrell encountered on their 1893 expedition from Edmonton to Baker Lake in Nunavut. In the area of Dubawnt Lake, about a day's canoe trip south of treeline, the Tyrrells discovered a landform that James described as a "moss glacier," an immense peat slope angling down gently from a ridge until it met the shores of Dubawnt Lake. At the lakefront, huge blocks of peat had fractured off the cut face of the moss glacier and had fallen like icebergs into the lake. Such permafrost-laden moss glaciers can be up to a mile and a half long and a mile or so wide. And along the cleft face of the glacier at the water's edge the chocolate brown, compressed peat may be up to seventy feet (21 m) thick. The Tyrrells' impression was that these immense peat slopes were moving like glaciers; that is, the pressure and weight of the accumulated peat was melting a deep permafrost layer, and the whole peat slope was slowly slipping toward the lake. The freshly cut, nut-brown face of the moss glacier revealed layer upon layer of peat moss—a fertile area for fossil hunters or those who wished to study changes in taiga vegetation over centuries.

Nor does the term *bog* adequately describe the palsas and peat plateaus that often develop in the taiga in areas of discontinuous or spotty permafrost. A palsa is a stretch of organic terrain 100 to 150 feet (30 to 45 m) across. Through the accumulation of peat and the progressive lifting effects of permafrost, a palsa can be elevated up to 25 feet (8 m) above the surrounding forest floor. Peat plateaus are even larger areas, but they form in the same way as palsas do, namely, from the slow accumulation of peat and the lifting, expanding action of ice. However, these miniature peat mesas are often scarred where the collapse of a tree has opened the forest canopy and allowed sunlight to penetrate, with the result that part of this peat plateau has thawed and collapsed back to ground level.

In many parts of the North and even as far south as Minnesota, a taiga explorer may encounter ribbed fens, a landform best appreciated from the air. Flying over a ribbed fen, one sees a gentle slope with numerous, elongated, water-filled trenches running parallel across it. These trenches are often no more than twenty feet (6 m) wide, but sometimes they are a mile (1.6 km) or more in length. Separating each trench from the next is a small elevated ridge, perhaps a yard (meter) or so in height, made out of a combination of sphagnum moss, sedges, stunted black spruce, and other muskeg vegetation. It is as if some giant—perhaps Wisahkecahk, the legendary Cree god-man—has ploughed furrow after furrow across the muskeg slope. An early scientific hypothesis suggests that an immense sheet of water moving down the very gentle slope indeed tore the continuous cover of the wetland vegetation, so that, over time, long parallel rips developed across it. While most boreal ecologists reject this tear hypothesis as too simplistic, the full explanation of how ribbed fens form is still elusive.

For these exotic, northern landforms, a more eclectic term is needed. Canadian scientists use an Indian word and refer to such lands collectively as *muskegs,* a term that includes all organic landforms. Muskeg is a Chippewa word meaning grassy bog, but according to N. W. Radforth, a Canadian ecologist, as a scientific term it is defined as "any tract of country showing a surface layer of living vegetation and a sub-layer of peat or fossilized plant detritus of any depth, which may be in any condition from water-saturated to quite dry." Living vegetation with peat underneath constitutes a muskeg. It is a landscape whose main features are determined primarily botanically and only secondarily by geological processes. Although muskegs form gradually, by a slow process of bioaccumulation, once started—be it a bog, fen, palsa, or peat plateau—muskeg dynamics are under way.

Drinking Your Weight in Water

When I studied sphagnum moss as part of an introductory botany course at university, the plant made little impression on me. At that point in my life, I had not hiked through purple-cushioned muskegs; I had not scrambled, soaking wet, out of a muskeg onto a dry, sandy esker and followed it for miles as it snaked over a vast northern wetland; nor had I admired sphagnum hummocks decked in red, chartreuse, and rusty orange. Those experiences created in me a desire to understand how this tiny moss creates such bizarre landscapes.

The dominance of sphagnum is based on its ability to absorb and retain water, and its water-holding capacity is due to several characteristics. First, the shoots and leaves of the plant are of such small dimensions and packed so closely together that a film of water envelops the complete plant. Second, that liquid coating extends even between plants. The all-pervasive moisture is due to capillary action: The closely packed sphagnum creates wicklike columns of moss, which effortlessly lift water to the top of a sphagnum hummock.

Even more striking is the microscopic anatomy of the sphagnum leaf itself. A mature sphagnum leaf is one cell thick. Under the microscope, a netlike pattern comes clearly into focus. The holes in the net are not empty spaces but hyaline cells—large cells that at maturity empty themselves of their cytoplasmic content and become water-storage facilities. The hyaline cells are connected to each other by well-developed pores, and thick spiral bands encircle the slender exterior of the hyaline cells so that their delicate cell membrane is supported when the cell is bloated with water. The solid part of the net is made of a different kind of cell—elongated, green, living cells that crisscross at various points to make up the interlocking cords of the net. These are the food-producing cells of sphagnum moss. They are capable of carrying on photosynthesis, that basic chemical process that spins air, water, and sunlight into a sugar called glucose, the basic energy currency used by almost all living organisms.

How much water can sphagnum soak up as a result of its microscopic, netlike structure? A lot! Scientists have documented that water-saturated sphagnum weighs seven and a half times as much as completely dry sphagnum. That's equivalent to an average-sized man drinking half a ton of water! Nor is this just simple absorption on the part of the moss; scientists have found that sphagnum carefully air dried and then resubmerged in water absorbs only 55 percent of its dry tissue weight. The fine microscopic structure of the hyaline cells, once it is disturbed, loses much of its ability to retain water.

The hyaline cells also provide the secret of sphagnum's almost effortless acidification of its environment. When I first explored muskegs and learned that sphagnum acidifies its surroundings, I assumed that the moss must synthesize some complex organic acid to achieve the effect. I also assumed that this strategy was an energy-expensive but effective way for sphagnum to kill off its plant competitors. However, through natural selection sphagnum has come up with a better idea. It has evolved a way of acidifying its surroundings that is the essence of simplicity and energy efficiency.

The inner surface of the hyaline cells carries out an ion exchange program. It has two important effects, and it works like this. The inner surfaces of the hyaline cells are lined with positively charged hydrogen ions, which are released whenever another positive ion bonds in its place. Valuable ions are absorbed from the surrounding water, and most often these are the positive ions of calcium, sodium, magnesium, or other nutrients required for plant growth. These nutrients are taken out of circulation, and simultaneously positive hydrogen ions are liberated, acidifying their surroundings. This ion exchange is sphagnum's knockout punch: It removes nutrients, acidifies the site, and inhibits decomposition. Yet sphagnum has expended little energy to carry out this process.

R. P. Bauman and J. Gully, the pioneer American ecologists working at the turn of the century, devised an intriguing experiment to test how sphagnum acidifies its environment. As G. B. Riggs describes, they theorized that if the moss synthesized its own organic acid, then the same amount of acid would appear whether sphagnum was submerged in distilled water or water with nutrients added. On the other hand, if their ion-exchange hypothesis was correct, then sphagnum would form very little acid in the distilled water but would quickly form a normal amount of acid in the nutrient-rich water. Their findings clearly supported the latter theory—positively charged nutrients are bonded onto the inner surface of the hyaline cells and hydrogen ions are liberated, acidifying the environment. To my mind, theirs was an insightful experiment about a truly ingenious plant.

Muskeg Dynamics

There is one more card that sphagnum plays as it breaks the food chain and ends up dominating its muskeg community. This adaptive tactic involves the depression of soil temperature and the formation of permafrost. The ecologist P. E. Heilman explains that sphagnum's microscopic structure gives it a high insulation value because, as they dry out, the countless microscopic spaces and drying hyaline cells become millions of tiny dead-air spaces. Keeping this point in mind, let's follow the moss through the seasons and see how it promotes the formation of permafrost.

During summer, the sphagnum on the surface of the muskeg tends to dry out, thus insulating the surface of the muskeg against the penetration of heat. The drier and hotter the weather; the thicker the dry layer of sphagnum becomes, and the thicker becomes the layer of insulation against solar heat. Sphagnum soils are therefore among the last soils in the boreal forest to thaw. The cold or frozen soils of bogs are an additional characteristic that has the net effect of bringing decomposition in these muskegs almost to a standstill. Often, autumn in the boreal forest is characterized by heavy rains, and winter starts with a number of freeze-thaw oscillations. Both send muskegs into winter in a water-saturated condition, with most hyaline cells, as well as many of the small spaces between the moss plants, filled with water. Water is an effective conductor of heat, so in its water-saturated state the insulating attributes of sphagnum are greatly reduced. Water-saturated peat conducts heat from deep in the sphagnum mass until a layer of frozen peat forms. In essence, sphagnum moss insulates against incoming heat and conducts outgoing heat. No wonder peatlands are growing points for permafrost throughout the entire taiga region.

Now let's consider the countertactics. Leatherleaf, Labrador tea, bog rosemary, to mention just a few plants, have evolved ways of coping with the chilly, undernourished soils of muskegs. For example, their leaves remain on the plant for several years and are evergreen throughout winter. These leathery, sturdy leaves cut down on the energy and nutrients that must be expended to start growth in the spring. In addition, there are many fine, hairlike structures on the undersurface of the leaves and a thick, waxy coat of cutin on the upper surfaces, which cut down on water loss. Much of the boreal forest experiences nearly continuous sunlight during spring and summer, and it is often a warm and drying sunlight. Yet peat soils are slow to thaw. These factors put plants growing in bogs under considerable water stress during the spring and early summer. Without such specialized leaves, these bog plants could not survive.

Insect-Eating Plants

One group of plants—the carnivorous plants—copes with the lack of nutrients available in bogs in a most interesting manner. Included in this group are the pitcher plant *(Sarracenia purpurea)*, the round-leaved sundew *(Drosera rotundifolia)*, and the common butterwort *(Pinguicula vulgaris)*, which are perhaps the most exotic members of our muskeg flora. All of them by one device or another trap insects and, through digestion (or at least decomposition) of these captured insects, supplement their diet with nitrogen and other nutrients, flourishing despite the impoverished conditions of sphagnum moss soils.

The pitcher plant is especially good at this game. Evolution has tucked and shaped its leaves so that they form what the northern Cree describe as *small kettles for carrying water*. Many northern natives, growing up with traplines or in fishing camps, have spent hours playing with these small red kettles, and a wry smile comes to their faces when they hear about scientists studying the throwaway toys of their youth. Nevertheless, pitcher plants are truly a fascinating product of evolution. The kettle leaves are colored a deep red, especially crimson around the rim and flowing down the walls of the kettle. The blood-red color is believed to attract flies and other meat-eating insects. The hairs inside the kettle point downward and are covered with a slippery gel. Those features invite insects to venture beyond the rim, usually resulting in a fall into the watery solution in the kettle, where they are trapped and finally drowned.

Some scientists believe that, to decompose the insects, the pitcher plant secretes digestive enzymes into this warm, Sun-heated solution. Other scientists claim that bacteria, microbes, and a mosquito larva specially adapted for living in pitcher plants are the agents of digestion and that the plant just benefits from the waste products of these animals. Whatever the mechanism—and it may vary among species of pitcher plant—the digested remains of insects supply these carnivorous plants with most of the nutrients they need for growth. Consequently, within the kettles of the pitcher plant, a tiny ecosystem has developed, containing plant, animal, and decomposer to support the pitcher plant in its growth. Darwin's phrase "Worlds within worlds within worlds" comes to mind.

Mining the Peat

Muskegs are not the tractor-swallowing wastelands that people think they are. They actually have important economic uses. Canada, with over half a million square miles (over 1,300,000 km^2) of muskeg, has more organic terrain than any other country (Russia, however, leads the world in muskeg research). The domestic and economic use of muskegs throughout their circumpolar sweep deserves a book unto itself; I touch upon just a few of these uses.

The earliest uses of sphagnum moss were developed by the Indians of the boreal forest, who found that certain kinds of sphagnum, if dried in the Sun and cleaned of twigs and other woody fragments, made wonderful diapering materials. Infants were

The hairy butterwort *(Pinquicula villosa)* supplements its diet by capturing ants, small flies, mosquitoes, and midges. When the sticky upper surface of its leaves captures an insect, its struggle stimulates the leaf to secrete an acidic fluid and enzymes. These slowly reduce the insect to a nitrogen-rich liquid, which is then absorbed by small glands on the leaf.

placed in a moss bag and thus were surrounded by dry, soft, insulating sphagnum. The moss bag was carried, and later, when the infant was a little bigger, it was tied onto a packboard. When the child grew older still, sphagnum was spread inside of its hide clothing. Not only do sphagnum diapers far outperform modern diapers in absorbency, but also the natural acids of sphagnum suppress bacterial growth—diaper rash was almost unheard of among children raised in moss diapers. Better yet, moss diapers were fully biodegradable: They could be disposed of almost anywhere without creating a garbage problem—a far cry from today's plastic-covered, disposable diapers.

The acidic and slightly antibiotic attributes of sphagnum made it a useful emergency wound-dressing material for soldiers in northern Europe during both world wars. This is valuable information for hikers, canoeists, and hunters to remember. Sphagnum moss has also long been a favorite with gardeners, who use it as a bedding material. Its acid, antibiotic qualities suppress fungal growth, and when the acid is neutralized, the sphagnum decomposes fairly quickly, providing a quick release of nutrients into the soil. These features have made the sale and export of sphagnum peat moss into an annual $130 million industry in Canada.

The food productivity of peatlands, especially of bogs on the southern edge of the boreal forest near large metropolitan areas, is appreciated by few. It is estimated that 100,000 acres (40,000 ha) of peatlands are under agricultural cultivation in Canada, mostly in Ontario and Quebec, and supply a good portion of the produce for the greater Toronto and Montreal areas. Once the peatland is drained and neutralized, vegetable crops can be planted. Leafy vegetables like lettuce, spinach, celery, cabbage, broccoli, and cauliflower do well, as do such root vegetables as onions, radishes, carrots, beets, turnips, and potatoes. It should also be pointed out that the color and characteristic flavor of fine Scotch and Irish whiskeys are produced from a carefully controlled burning of blocks of sphagnum peat. Chivas Regal may be aged for twelve years, but it's a seedy old muskeg that gives it its distinctive taste.

In British Columbia, Michigan, Wisconsin, and New England cultivated bogs support a multimillion-dollar cranberry industry, and in the northern Prairie Provinces, open-water areas of bogs and fens have been developed into major wild rice production areas. In northern Saskatchewan, the development of a wild rice industry has greatly enhanced the area's economic prospects. Other aquaculture enterprises, such as trout farming, may also be developed in the abundant bogs and small lakes of the boreal forest.

Sphagnum also has considerable potential as an industrial filtering material. The use of dried blocks of peat as a home-heating fuel in Ireland is well known, but few are aware that Russia uses peat as a biofuel for generating electricity and that Sweden uses it to fuel its district heating and power stations. Furthermore, Russia has spearheaded a new use of sphagnum moss, drying it and pressing it into construction boards with high insulation value. These sphagnum construction panels are widely used throughout the taiga regions of Russia. Closer to home, a research group called the Natural Resources Research Council, located in Duluth, Minnesota, has experimented with peat as a basis for a petrochemical industry. So far they have been able to extract a number of economically useful chemicals from the distillation of sphagnum peat moss. If their research proves economically viable, we may in the future be driving our cars on liquefied or gasified sphagnum peat.

Wild rice growing in a gently flowing stream on the Canadian Shield.

Tally these economic uses of wetlands, add the value of muskegs as water reserves and as habitat for waterfowl, moose, and other wildlife, and it becomes clear that, far from being wastelands, muskegs are valuable natural regions that deserve careful management and protection.

There is one more benefit we all derive from muskegs—one that we are often unaware of. The taiga contains one of the largest deposits on Earth of immobilized organic carbon—nearly 830 gigatonnes, a vast majority of it in the boreal peatlands. Thus since the end of the last ice age, with the expansion of its muskegs the boreal forest has been acting as a huge circumpolar sink, removing carbon from the atmosphere and storing it in its peat deposits. The value of this carbon sink has greatly increased as industrialized societies burn fossil fuels at an accelerating rate, releasing significant amounts of CO_2 and other greenhouse gases into the atmosphere. The carbon sink of the boreal forest has offset global climate warming by removing 0.22 tons (0.2 tonne) of carbon from the atmosphere each year and immobilizing it as accumulated peat deposits.

However, these gigantic "lungs" of the circumpolar boreal forest have not been keeping up, and some scientists believe that the accelerating consumption of fossil fuels is warming our climate at a rate of 4°F (2°C) approximately every twenty-five years. As I review in Chapter 2, this increase in temperature might not sound like much, but in the boreal forest this gradual warming is believed to be drying out large areas of forest, dropping the water table, accelerating the rate of decomposition, and greatly increasing the frequency of forest fires. These effects could change the circumpolar taiga from a carbon sink into a carbon source. As the immense carbon reserves of the taiga become mobilized, the addition of carbon dioxide, methane, and other greenhouse gases to the atmosphere could significantly accelerate global climate warming.

As previously mentioned, scientists from many disciplines are investigating the extent and implications of these interactions. For example, they are concerned about a possible increase in forest fires worldwide, a drying out of agriculturally productive regions, a melting of glaciers, with a possible flooding of coastal cities. Researching each of these possible effects and what can be done about them will continue to be important into the future. Two of the questions such research needs to address are, Should industrialized societies impose a carbon tax to lower the use of fossil fuels? Or can these societies convert to alternative and nonpolluting energy sources, like solar power and hydrogen gas? These are important questions, and maintaining the boreal forest as a circumpolar carbon sink will be a crucial part of these deliberations.

The Wihtiko, Tollund Man, and the Power of the Muskeg

No discussion of muskegs would be complete without mentioning its most infamous inhabitant, the Wihtiko (plural, Wihtikowan), the cannibalistic being of northern native mythology. Wetiko, Whitagoo, and Windigo are alternative names, but I use the Woods Cree word.

I first encountered a Wihtiko in a painting in the Government Building of La Ronge in northern Saskatchewan. The painting is by a local aboriginal artist, John Halkett, and the inscription next to it was written by a local Woods Cree person. It reads:

> In times when food was scarce many people died of starvation, but there were those who resorted to cannibalism and were transformed into Wihtikowan. These creatures were miserable and frightening. Their lips were chewed, their clothes were tattered, and they had a large block of ice in their body. When a Wihtiko traveled, usually alone, his heartbeat could be heard for miles away. Medicine men could sometimes see where he was in their dreams. In spite of these warnings, the Wihtiko was a serious threat. This supernatural man-eater was immensely strong and difficult to kill. Once defeated, he was usually burned to death, but this took two or three days of burning because of the ice in his body. It was said by Christianized Indians that a page from the Bible used as gun wadding could kill a Wihtiko.

Wihtikowan are the creation of the northern Algonquin people—the Cree, the Montagnais-Naskapi, the Ojibwa, and the Saulteaux. The first written accounts of Wihtikowan, as early as 1600, are in the journals of European explorers, but their origins undoubtedly extend into prehistory. Wihtikowan have wide geographic roots; sightings having been reported in the cultures of the Passamaquoddy, the Micmac, and the Iroquois peoples. A Wihtiko was the ultimate horror: once a hunter of game, now a hunter of people. It did not remember that it was once a human being. Starvation turned it into a Wihtiko, and its death was the only known solution.

The Wihtikowan capture the spirit of the taiga. For people living off the land in the taiga, it doesn't take long to realize that periods of food shortage are the rule, not the exception. Food usually comes in large packages—an adult moose, for example, or a herd of caribou. However, these mother lodes of protein are infrequently encountered, and much traveling and many hungry days might intervene between successful hunts. Taiga lakes are also inconsistent producers of food; some have fish in abundance, others do not. In the taiga, living off the land led to a demanding, nomadic lifestyle. The sparsity of prey and other food resources dictated that only a few families at most could travel together and expect to find sufficient food. Such circumstances helped generate a powerful taboo on cannibalism and ultimately led to the birth of the Wihtiko.

Wihtikowan preferred winter to summer because, in winter, people traveled and hunted farther afield and thus were easier to capture. Wihtikowan hid in muskegs during summer, sleeping for long periods, and using the permafrost and the frigid muskeg soils to prevent their ice cores from melting. Bill Merasty, a member of the Peter Ballantyne First Nation from northern Saskatchewan, describes what an encounter with a Wihtiko was like:

> A wrestling match with a Wetiko was no ordinary scuffle. As man and Wetiko struggled, the ferocity of the battle made them rise to the treetops. If the man were able to yell louder than the Wetiko, the cannibal would fall to the ground and the man could kill it. It was a shouting match in which the loser was, in a sense, beaten by sound. Victory came only to someone with strong medicine power, someone who could not lose.

Perhaps the Wihtiko legend was partially fueled by the remarkable preservation of human remains buried in muskegs. The ability of sphagnum peat to preserve skin, hair, and cloth is derived from its waterlogged, cold, acidic conditions. In 1925 in the sphagnum peatlands of southern Sweden a large woven cloak was recovered so perfectly preserved that the details of its weaving and even the tension lines from its having been worn were visible, even though the cloak dated back to the Late Bronze Age. Throughout the bog areas of Scandinavia, axe handles, cart rails, and other remnants of long-forgotten cultures have turned up in remarkably well-preserved condition.

The remains of a considerable number of bog-buried human beings have been recovered from Danish and German peat bogs. According to H. Godwin, the most remarkable is Tollund Man. K. Thorvildsen, who unearthed Tollund Man in 1950, showed through various forms of evidence that Tollund Man lived approximately 1,000 years ago. His skin, nails, hair, and fur-lined cap were intact, as was the plaited rope encircling his neck—which had been used to strangle him. So well preserved was his stomach that the fruits and seeds of the porridge that he ate a millennium ago were all recognizable.

Muskegs are an unusual weave of themes: perfectly preserved corpses and alleged cannibalistic hunters, landscapes created by a three-inch-high primitive moss, and plants that dine on insects. The peculiar colors of muskegs—the luminous purples, crimsons, and chartreuses of sphagnum moss—add to their extraterrestrial quality. To be lost in a muskeg near sundown, surrounded by gnarled tamarack trees festooned with lichens that conjure up the ghosts of Jacob Marley and his fellow travelers; to fight off hordes of flies and mosquitoes while searching for a dry piece of land upon which to spend the night—these are experiences that leave a deep impression on a person. No, I am not sure that I can say with complete confidence that in the vast, ageless muskegs of the North—a land shaped by a primitive plant that fits comfortably in the palm of my hand—there is nothing to dread.

General References

Godwin, H. 1981. *The Archives of the Peat Bogs*. Cambridge: Cambridge University Press.

Gore, A. J. P. 1983. Mires: Swamp, Bog, Fen, and Moor. In *Ecosystems of the World*, vol. 4A, edited by A. J. P. Gore. Amsterdam: Elsevier Scientific Publishing.

Heilman, P. E. 1967. Relationships of Availability of Phosphorus and Cations to Forest Succession and Bog Formation in Interior Alaska. *Ecology* 49:331–36.

Heinselman, M. L. 1963. Forest Sites, Bog Processes, and Peatland Types in the Glacial Lake Agassiz Region, Minnesota. *Ecological Monograph* 33:327–74.

Johnson, D. L. Kershaw, A. MacKinnon, and J. Pojar. 1998. *Plants of the Rocky Mountains*. Edmonton: Lone Pine Publishing.

Larsen, J. A. 1982. *Ecology of the Northern Lowland Bogs and Conifer Forests*. New York: Academic Press.

Merasty, M. 1974. *The World of Wetiko: Tales from the Woodland Cree*. Saskatchewan: Saskatchewan Indian Cultural College, Curriculum Studies and Research.

Riggs, G. B. 1940. The Development of Sphagnum Bogs in North America. *Botanical Review* 6:666–93.

Tyrrell, J. W. 1908. *Across the Sub-Arctics of Canada*. Toronto: Briggs.

Van Cleve, K., F. S. Chapin III, P. W. Flanagan, L. A. Viereck, and C. T. Dyrness, eds. 1986. *Forest Ecosystems in the Alaskan Taiga: A Synthesis of Structure and Function*. New York: Springer-Verlag.

Watts, A. T. 1957. *Reading the Landscape*. New York: Macmillan.

Zoltai, S. C., and F. C. Pollett. 1983. "Wetlands in Canada: Their Classification, Distribution, and Use." In *Ecosystems of the World*, vol. 4A, edited by A. J. P. Gore. Amsterdam: Elsevier Scientific Publishing.

10
Northern Lakes, Troubled Waters

A fond memory from my early years in the taiga is of an evening spent on a cobblestone beach of a northern lake with a Metis trapper, barbecuing a lake trout. Ben, who had lived in this region of northern Alberta all his life, had caught the mammoth trout the day before. As the fish cooked slowly on low embers we talked about such diverse topics as whether the peculiar bumps that grow on the lips of black bears might function as a berry rake for gleaning raspberries and blueberries more easily from their bushes; how to collect and prepare birch syrup in the spring; and what loons tasted like and how best to cook them.

Much trout and beer were enjoyed that night, as we watched the northern lights flicker and dance overhead. Ben told me that you should never whistle at them. The Northern Cree believe that the northern lights are their ancestors dancing in the sky; if you whistle, they will come down and take you up dancing with them, never to return.

Another memory is of dipping my cupped hands into any lake or stream and drinking the clean, pure water. I never got sick; I never worried about drinking unclean water. Today I drink water straight from boreal lakes and rivers only cautiously, if at all. The luxury of drinking directly from lakes and rivers, which was available almost everywhere in the Canadian taiga until the early 1970s, now exists only in the far North. Today a water filter and treatment kit—standard equipment for all hikers, canoeists, and hunters—accompany me on all expeditions unless I am very far north. Thirty years ago we never packed them.

This change is expressive of the serious degradation that boreal lakes, rivers, and streams are undergoing all across Canada. A large part of this change is due to the invasion of *Giardia*—an intestinal protozoa that has become established in many boreal lakes and rivers, often transported there through human wastes. *Giardia* causes "beaver fever" in humans, whose symptoms are nausea, diarrhea, and dehydration and that warrants the attention of a physician as soon as it develops. Other serious concerns about water quality in the taiga are also due to human-caused contaminants—mercury, pesticides, and sulfuric and nitric acids—which increasingly affect the ecological integrity of water bodies in the boreal forest. To understand these problems and their causes better, we must first examine some of the unique properties and processes of boreal lakes and rivers. Let's start by examining the fundamental differences between land and water environments in the taiga.

Water Habitats versus Land Habitats

In the taiga, lakes are characterized by temperature stability while land habitats are characterized by temperature fluctuation. In northern Saskatchewan, for example, during the summer the ambient temperature in the forest can get as high as 90°F (40°C), while during the winter the mercury can drop close to −60°F (−50°C). Between January and July, that's a difference of 150°F (90°C)—quite a spread, and the animals and plants of the forest expend considerable effort and evolutionary ingenuity coping with such seasonal fluctuations in temperature. In the lakes of this region, on the other hand, the temperature of the water seldom rises above 70°F (20°C) even during the dog-days of August, while throughout the winter the temperature of the lake water under the ice is always slightly above freezing. From January to July, aquatic plants and animals experience a change in water temperature of only about 40°F (20°C), or about a quarter of the temperature range of terrestrial environments. Furthermore, plants and animals that live under the ice of lakes and rivers are never exposed to freezing temperatures. The temperature stability of lakes is one of the great assets of an aquatic lifestyle.

On the other hand, water conducts heat more than twenty times as effectively as dry air. This physical attribute of water has many ecological effects (we looked at some of them when discussing the insulating characteristics of snow and sphagnum moss). This heat-conducting capacity of water has caused many birds and mammals to evolve "waterproof" feathers and fur to reduce heat loss. The long guard hairs and extensive undercoat of the beaver's pelage, for example, maintain a layer of dry air next to the skin even while the animal is swimming. The beaver spreads oil from glands at the base of its tail to lubricate and further waterproof its "dry suit."

An aquatic existence offers other challenges. Take, for example, oxygen: O_2 is plentiful in the air—it makes up 21 percent of Earth's atmosphere and is reasonably evenly distributed over its surface. However, the amount of oxygen gas that will dissolve in water before O_2 becomes saturated and bubbles out, returning to the air, is quite restricted, making oxygen a scarce resource in a water environment. A few conditions can affect it. For example, the colder the water, the more oxygen the water can hold. That means that cold northern lakes are relatively oxygen rich. But even in these lakes there are only twelve molecules of oxygen gas for every million molecules of water. Because oxygen is so scarce even in northern lakes, understanding the distribution of oxygen gas dissolved in a lake and how it is renewed is key to understanding the dynamics of a healthy lake.

The scarcity of oxygen in water explains why warm-blooded animals, if they take on an aquatic lifestyle, never abandon their habit of returning to the surface of the water to breathe. The river otter *(Lutra canadensis)*, sea otter *(Enhydra lutris)*, loon *(Gavia* species), beaver, muskrat, sea lion, seal (Pinnipedia), whale, and porpoise (Cetacea)—although they might be completely adapted to making their livelihood under the surface of a lake or ocean—always return to the surface to breathe. Water simply does not contain enough oxygen to support the high metabolism of warm-blooded animals

The Turnover of Lake Water

If oxygen dissolved in the water of a lake is such a rare commodity, and if all organisms in the lake use it, why is it not depleted? How is oxygen in lake water renewed? There are actually two processes that replenish oxygen in lake water. One process is photosynthesis, carried out by aquatic plants. The algae floating in the water as well as the rooted, floating plants growing along the shoreline—by taking in carbon dioxide, water, and energy from sunlight and turning that into glucose (the basic energy source for all living cells) and then giving off oxygen as a waste product—help to renew the oxygen dissolved in a lake.

The other process that renews oxygen in lake water is the physical turning over of water in the lake twice each year, once in the spring and once in the fall. To understand this process, we must remember that water is at its densest condition at 39°F (4°C). That means that more molecules of water fit into a cubic inch of space at this temperature than at any other. When water cools below this temperature, water molecules begin to link up in the orderly crystalline structure of ice. Water molecules bonding together in this crystal pattern take up more room, and water becomes less dense. As a result, ice floats. On the other hand, if water is warmed above 39°F (4°C), heat causes the water molecules to vibrate more vigorously against one another, with the result that the water molecules take up more and more room as their temperature rises.

These physical attributes of the water molecule affect how a lake functions. One of the most important effects is that ice floats. If ice were more dense than water, our lakes and rivers would freeze from the bottom upward, and fish and aquatic plants would be frozen in solid ice for seven months of the boreal winter. Many chemicals are denser in their frozen state, but not H_2O. It is fortunate that ice floats, which means that boreal lakes and most northern streams and rivers are capped with a protective layer of ice during the harsh taiga winter.

Another important ecological process determined by the physical properties of the water molecule is lake turnover, a process essential to the health of a lake. When after six or seven months of winter the ice disappears, the lake water is once again exposed to the air. Through the random exchange of gases, carbon dioxide, ammonium, methane, and other gases are given off to the air and oxygen gas is replenished in the lake water. The increasingly long days of spring heat the water on the surface of the lake until it is warmed to 39°F (4°C). At this temperature, surface water is at its densest, and this water actually sinks, pushing the deeper water of the lake up to the surface and exposing it to the air. Again, carbon dioxide, ammonium, methane, and other gases are released, and oxygen is taken up by this water until O_2 reaches its saturation point at approximately twelve parts per million.

As a result of this dynamic, a huge, slow convection current is established in the lake during the spring: The surface water is warmed by solar radiation to 39°F (4°C), then this denser water sinks and pushes water from the depths closer to the surface. This convection current continues until all the water in the lake takes on the same temperature (39°F or 4°C). Now even slight winds cause deep stirring of the lake. All the water in the lake is eventually exposed to the air, and oxygen is renewed throughout. This turnover of the lake also stirs the bottom sediments, and if you observe closely you can often see the water of the lake turn slightly browner or darker from the distribution of silt and nutrients.

Turnover continues until a warm layer of water becomes established on the surface of the lake. This layer of warm water—the one that all northern swimmers try to stay in—has a lesser density than the water beneath it, and if it becomes solidly established the warm water on the surface seals off the deeper water from contact with the air just as effectively as ice does in winter. Strong summer winds or cold rains may disrupt the warm layer, but as long as it is established, the deeper water is not exposed to the air, and only photosynthesis can renew its supply of oxygen. A lake in this condition is said to be thermally stratified, which has important ecological effects. For example, many northern fish species (e.g., lake trout, lake whitefish, and walleye) prefer cool water, and they move into the cooler, deeper parts of a lake during the summer. However, if a lake stratifies early, the oxygen reserves of this deep, cool water can become depleted by late summer.

In small boreal lakes, however, during a hot, sunny spring, ice can disappear and, in the same day, thermal stratification can become established, resulting in an incomplete spring turnover. An incomplete autumn turnover can result from early fall snowstorms, which can cool lakes rapidly, often with ice forming soon after. Normally boreal lakes cool, destratify, and turn over for several weeks before their surface freezes. These long autumn turnovers are more frequent on large boreal lakes (with their larger thermal capacities) than on smaller lakes, which respond more quickly to colder weather. These healthy autumn turnovers eliminate the accumulation of harmful gases in the lake water and renew the oxygen reserves throughout the entire lake before the ice cover seals the lake from contact with the air for six or seven months.

The Passage of Ice

Most boreal lakes freeze over during October or early November, and the ice that forms can be black or white or a mosaic of the two. Transparent black ice forms when water loses enough heat that the surface of the water freezes. It is black because the dark lake water and lake bed beneath is visible through it. Black ice, being transparent, allows sunlight to shine through it, so photosynthesis can occur in the lake water beneath the ice—albeit at a reduced rate compared to summer. Alternatively, boreal lakes may freeze over with white ice, ice that incorporates snow, air bubbles, or ice particles (frazil ice) as it forms. This type of ice appears white because it reflects sunlight. If a northern lake freezes over with white ice, then the underlying water remains in relative darkness all winter; little photosynthesis takes place, and no oxygen is renewed.

Snow cover is also important. If the winter brings thick snow cover, the water under the ice remains in darkness, and little photosynthesis occurs. On the other hand, if strong winds periodically blow the surface of the ice free of snow, and black ice covers at least portions of the lake, photosynthesis begins again, and oxygen is partly renewed. Strong winter winds, which are such a threat to animals living on the land, are thus a benefit to the animals living in a lake. They sweep lakes clear of snow and allow oxygen to be renewed, which can help prevent a winter kill of fish.

The spring breakup of the winter ice completes the annual cycle. Breakup on lakes differs from breakup on rivers. The first ice-free water on a lake usually appears in a moat around its outside edge or where a strong current of water enters or leaves the lake. These areas of open water provide important microhabitat for loons, swans, geese, ducks, and other water birds that migrate north during early spring.

Lake ice melts because of a strong spring Sun pouring down on it during days of increasing length and because of solar radiation absorbed by the crystalline structure of the ice. During the last stages of breakup, the ice weakens and breaks into millions of vertical cylinders of ice packed closely together. These ice candles are approximately one inch (3 cm) thick and can be ten or twelve inches (25–30 cm) high. A gentle wind blowing through these cylindrical ice crystals creates a special music — the funeral knell of winter. When I sit on the shore of a boreal lake on a May evening and listen to the sound of a million ice candles blending with the distant cry of geese and sandhill cranes, only then do I truly believe that a gentler season has at last arrived.

Ice on the Northern Rivers

Breakup of northern rivers occurs in a different way from that of the boreal lakes. Icemelt occurs rapidly in the taiga in the spring, and the flow of water over the land and through creeks and streams is at its greatest. Oftentimes rivers clear their ice without incident, and the meltwater from winter's snow cover simply drains away. Other times, spectacular ice jams occur, and the buildup of ice and water behind these ice dams can have major ecological and human impacts. What causes ice jamming? Because rivers in the North usually break up before large lakes shed their cover of ice, ice jams can occur where one of these rivers enters a large lake. Furthermore, across boreal Canada as well as Siberia, many major rivers flow northward. On the southern portion of these rivers, the thaw occurs; rivers break up, and water and broken ice flow northward to where the river is still frozen. It is at that point that massive ice dams can form.

Conditions such as these occur on the Mackenzie River, which flows out of Great Slave Lake on its way to the Arctic Ocean. One summer evening, as I was admiring the view from the high banks of the river at the edge of the town of Fort Simpson, in the Northwest Territories, a local resident strolled by and told me that the town had been almost flooded that spring by a huge ice dam that had formed twenty-five miles (40 km) downriver. Day after day townspeople had helplessly watched the river rise foot by foot up those massive seventy-five-foot (23 m) banks. Thankfully, the ice dam broke, and the river dropped, before the town had to be evacuated.

River ice dams and the flooding that results can be beneficial processes, too. The silt from floodwater fertilizes riparian forests; floodwater recharges ponds and small lakes perched on terraces, located next to these northern rivers; and floodwater rejuvenates the productivity of deltas. Consider, for example, the ecological role of flooding in the Peace-Athabasca delta. In northern Alberta, the large Peace and Athabasca Rivers come together and form a delta that empties into the west end of

Lake Athabasca. The Peace-Athabasca delta, one of the largest inland, or freshwater, deltas in the world, has been developing for more than 10,000 years, and it is an impressive and productive northern ecosystem. This delta, 80 percent of which is located within Wood Buffalo National Park, is home to and breeding grounds of approximately 215 species of birds, 44 species of mammals, and 18 species of fish. When its water levels were at their peak some twenty-five years ago, more than 600,000 waterfowl hatched on the Peace-Athabasca delta every year.

Flooding renews the productivity of the Peace-Athabasca delta. Spring floods are caused by the ice dams that form in various parts of the delta. The silt from the flood fertilizes large sedge meadows, and the water from the flood recharges ponds and "perched basins" located next to the river. These small water bodies are habitat for waterfowl, muskrats, moose, and other animals. However, not all the effects of the ice dams are positive. For example, spring floods during the late 1960s and early 1970s inundated and drowned 5,000 bison *(Bison bison)*, cutting the bison population of Wood Buffalo National Park nearly in half.

Man-Made Dams

In 1968 British Columbia completed the construction of the W. A. C. Bennett Dam on the Peace River in central northern British Columbia. The Bennett Dam reduces the flow of the Peace River during spring to the point that ice jams no longer occur in the Peace-Athabasca delta. As a result, these perched basins are no longer rejuvenated; large areas have changed from productive sedge meadows into stands of aging willows, and there have been enormous declines in the populations of fish, waterfowl, and fur-bearing animals. Productive bison habitat in the delta has been reduced to such an extent that the bison are restricted in their movements and predictable in their feeding habits, with the result that wolves now prey upon and remove a majority of the calves born into these herds each year. Furthermore, the Cree, Chipewyan, and Metis groups that depended upon the Peace-Athabasca delta for their livelihood have suffered severe economic hardship. The Bennett Dam has reduced one of the most productive delta systems of the world to a crippled and broken ecosystem.

Dams cause other major problems across the taiga for the wildlife and humans that live in the boreal forest. This subject involves many complex problems and interactions that can only be briefly discussed here. David Schindler and others offer good reviews of some of the controversies and significant environmental impacts associated with hydroelectric dams.

In the taiga region of Canada, there is a long tradition of huge water-diversion schemes and massive hydroelectric dams. Large reservoirs have been constructed so that the water pouring through these dams can drive giant electrical generators. Large hydroelectric dams are located in the boreal regions of most of the Canadian provinces—for example, the Churchill Falls project in Labrador, the James Bay development in Quebec, the Nelson River dams in Manitoba, the Tobin Lake Dam in Saskatchewan, and the W. A. C. Bennett Dam in British Columbia.

The former premier of Quebec, Robert Bourassa, championed the James Bay hydroelectric developments based on the principle that these dams were generating "good, clean energy." We are learning otherwise. Destruction of beaches and shorelines, devastation of beaver and muskrat populations and their predators, drowning of caribou and destruction of their migration routes, and decline of fish populations are some of the well-documented impacts of these "good, clean" boreal developments. It has also been shown that hydroelectric reservoirs can emit significant quantities of greenhouse gases following the flooding of wetlands and peatlands.

We have also discovered during the past two decades that flooding wetlands creates a significant source of mercury contamination. When wetland vegetation decays under oxygen-poor conditions, mercury atoms are released from the soil and from decaying vegetation. More serious yet, these mercury atoms are transformed by bacteria into methyl mercury, a form that is readily absorbed by plants and animals. David Schindler, a leading aquatic ecologist in Canada, states: "In almost every case, raised water levels have caused fish populations to become highly contaminated with mercury, frequently exceeding limits for commercial marketing." Not surprisingly, mercury contamination is high in many aboriginal communities across the North. From eating fish—a staple in the diet of northern, isolated communities—many of these people have blood mercury levels approaching the range at which neurological symptoms could occur, although no such effects have yet been documented. It may be especially important to determine the effects of this mercury contamination on newborn babies, but this type of study has not yet been carried out in any northern community.

Lake Baikal

The taiga is unquestionably the land of lakes. Freshwater boreal lakes, underlain by ancient igneous rock formations, are the most numerous kind of lake in the world, probably numbering more than 2 million worldwide. Seen from a spacecraft circling the Earth, boreal lakes shine like jewels on the evergreen mantle that cloaks the shoulders of Mother Earth. These boreal lakes are an irreplaceable resource, containing more than 80 percent of the world's unfrozen fresh water. They range in size from small ponds and kettle lakes hidden away in the countless corners of the boreal forest to the largest boreal lake of them all, Lake Baikal, the sacred sea of central Siberia and to my mind one of the natural wonders of the world.

Lake Baikal is nearly 400 miles (635 km) long and averages 35 miles (56 km) wide, and its deepest confirmed point is 5,314 feet (1,620 m) of water. Lake Baikal by itself contains 18 percent of Earth's fresh water—15,000 cubic miles (32,000 m^3) of it. It is not only one of the largest and deepest lakes on Earth; it is also one of the clearest. Its clarity is largely due to its depth, because even the most violent windstorms do not stir its bottom sediments. Lake Baikal has an unearthly clarity, which makes the lake appear a deep blue, irrespective of the weather. During May, the clearest season, sunlight penetrates the water to a depth of 130 feet (40 m). In most lakes, if sunlight penetrates 25 feet (16 m), the water is considered exceptionally clear.

Lake Baikal is also the oldest known large lake in the world; it has been in its present form for at least 23 million years. If the lake is a geological wonder, it is a biological wonder as well. Because of its great age and geographic isolation from seas and other lake basins for much of its history, a unique flora and fauna has come to occupy the lake. Of its 500 distinct plant forms and over 1,200 animal species, more than two-thirds are found nowhere else on Earth. For example, nearly all of its arthropods species and over 80 percent of its fish species are unique to the region. The Baikal sturgeon is one of these endemic fish species. Adult Baikal sturgeon grow to as much as seven feet (2.1 m) in length and weigh over 500 pounds (230 kg), producing not only delicious of meat but, in season, extremely valuable black caviar. The Baikal sturgeon was overharvested during the first part of the 1900s, but Russians have built large hatcheries in order to restock the Baikal sturgeon not only into Lake Baikal but also into other boreal lakes and reservoirs across Siberia.

Another fascinating creature is the Baikal seal *(Phoca sibirica)*, or *nerpa*, as it is called locally. Lake Baikal is one of the few freshwater lakes in the world that supports a large population of seals year-round. Nerpas feed on the abundant fish of the lake, and because of its depth Lake Baikal does not freeze over until some time in January. After freeze up, the Baikal seals live in open leads and in the tributaries leading into the lake until the lake breaks up during May and early June. At present, nerpas are estimated to number approximately 25,000.

Public controversies over the use and harvesting of Lake Baikal's resources have erupted in Russian society several times during the past decades, but public support for the protection of Lake Baikal remains solid among Russians.

The Shield-Edge Lakes of the Canadian Taiga

In Canada some of the largest lakes of the taiga can be described as shield-edge lakes, lakes that form only where the Precambrian shield joins other geological regions, such as the Interior Plains. Shield-edge lakes mark the southern boundary of the Canadian Shield, the core of the North American continent. They form on a line that cuts across the North American continent from the Great Lakes in the south to the Arctic Ocean in the North. Lake Superior, Lake of the Woods, Lake Winnipeg, Lake Athabasca, Great Slave Lake, and Great Bear Lake are the largest and best known of these lakes, but there are many others marking the southern border of the Canadian Shield. They are truly the great lakes of Canada's taiga. They are more productive than normal boreal lakes because they contain features of both the sedimentary rock of the Interior Plains and the igneous and metamorphic rock of the Canadian Shield. Thus these lakes have the great depth and clear, cold waters typical of oligotrophic Precambrian shield lakes, and yet they have the higher nutrients and richer food resources of lakes of the Interior Plains.

Because these lakes are among the most productive boreal lakes in North America, they should be the focus of careful research and effective management. Thus far Canada has not met these needs well. In a recent interview with me, David Schindler, of the University of Alberta, said,

> I've made the case several times over the years that our lack of information about Great Slave Lake, Great Bear Lake, and the other taiga great lakes is a national disgrace. More Canadian money goes to studying African great lakes than our own northern ones. There are good assessments on Great Slave Lake from the 1940s, thanks to D. S. Rawson, but very little research has been carried out since then, and almost no research has been carried out on Great Bear Lake.

Shield-edge lakes deserve the highest degree of protection. I hope that Canada will choose to provide proper ecosystem management for these great lakes of the taiga before it is too late.

Boreal lakes show some of the same characteristics that define the northern forest, containing, for example, fewer species than freshwater lakes farther south. Researchers at the Experimental Lakes Area in northwestern Ontario, an internationally known research center, calculate that the shield lakes of their region contain no more than a few hundred species, a number that includes all plants, invertebrates, and fish. Small boreal lakes often contain only one predatory fish species at the top of the food chain—the northern pike—and large boreal lakes add only a few more species as predators at the top of their food web—lake trout, cisco, lake whitefish, walleye, and dolly varden.

Animals from northern lakes like it cold. Technically they are known as cold stenotherms, species that can survive only in a narrow range of cold temperatures. Most boreal fish species and aquatic invertebrates are unable to survive in warm water or water with low reserves of dissolved oxygen. These traits are expressive of their long association with glacial lakes. These species survive the summer by migrating into the deeper, colder waters of the lake and migrating upward as the lake cools during autumn. One of the concerns about global climate warming is that it will reduce cold-water habitat in many small and medium-sized boreal lakes and threaten the populations of cold stenothermic species.

Another attribute that many boreal lakes show is their short growing season and thus their limited productivity. Research shows that it takes a surprisingly large area of the lake—usually 1,200 to 12,000 square yards (1,000 m^2 to 10,000 m^2) of lake surface—to support a single lake trout, walleye, or other large predatory fish. Working that out, it means that many boreal lakes contain only between 200 and 500 fish that are worthy of being caught. This is not a very large population for a summer's fishing. A family out for a day of fishing who has caught 10 nice-sized fish may be removing between 2 and 5 percent of the large fish that live in that lake.

Fish populations in boreal lakes are unproductive for other reasons as well. Typically, lake trout, northern pike, and walleye in cold boreal lakes do not reproduce before they are four to six years of age. These fish reproduce every year, but survival of their fertilized eggs and young fish is a much rarer event, typically occurring only once every six to ten years. In these cold lakes of the North fish can grow extremely slowly and survive for twenty to thirty or more years. After one of these fish reaches maturity, much of the year's energy intake is funneled into reproduction. This reproductive effort reduces growth to the extent that, until recently, these fish could not be accurately aged, leading to gross overestimations of the sustainable harvest of these fish populations.

Seemingly innocuous fishing habits can have a major impact on boreal lakes. Dumping a bucket of live bait into the lake at the end of a day of fishing seems a simple act but it can introduce new species of minnows or invertebrates into boreal lakes. Bait buckets should be carefully drained onto sunny, sandy ground, where there is little chance of this water draining into a creek or lake. Minnows, fish eggs, or other organisms used as bait should be disposed of in bear-proof garbage containers. The use of lead sinkers is another fishing practice that should be avoided. Lead weights and sinkers that become entangled and are left behind in a lake are eaten by loons, ducks, and other waterbirds, which mistake them for small prey or for the pebbles they use to grind food in their gizzards. Lead poisoning is a significant cause of mortality among these birds. Research in New England and New York found that 14 of the 222 loons in their study died from lead poisoning; 11 of the 14 had lead sinkers in their stomachs. Fishing equipment made of lead has been banned in Canadian national parks; anglers should use only steel or organic weights and sinkers, which are nontoxic and will decompose over time.

Endangered Waters, Threatened Forests

On a worldwide basis, one of the greatest threats to the ecological health of boreal lakes and rivers is acid precipitation. It is a problem important in its own right, and it is a problem that shows how inexorably linked terrestrial and aquatic ecosystems are. How do our life-giving rain and snow, and even fog, turn acidic?

Acid rain forms primarily from emissions of sulfur dioxide (SO_2) and nitrogen oxide (NO_x). When these pollutants are mixed with water vapor in the air, they spontaneously transform into sulfuric acid (H_2SO_4) and nitric acid (HNO_3)—molecule for molecule, some of the most powerful acids in existence. These acid-making pollutants can be transported by winds for hundreds of miles before they settle out on plants, lakes, buildings, and the surface of the Earth. Thus the smokestacks and car exhausts of places like Chicago, Detroit, and Toronto are creating the acid rain that is killing the lakes and northern forests in Quebec, Labrador, and Newfoundland.

Extensive areas of the taiga—particularly Precambrian shield areas—are highly sensitive to acid rain. Precambrian shield areas are found not only across northern Canada but also across Scandinavia and eastern Siberia. These vast areas of granitic bedrock contain very little limestone in their soils and lake beds that can neutralize acid precipitation. As a result, the lakes of shield areas turn acidic fairly quickly, with dire ecological results. For example, hundreds of lakes in the Canadian Shield area of Ontario are already devoid of fish because of acidification. In Quebec more than 1,300 lakes are currently acid stressed and in danger of losing all fish populations. In Sweden hundreds of lakes are in a similar condition. However, the effects of acid rain are not limited to Precambrian areas. Some experts estimate that over 60 percent of the trees growing in Germany show reduced growth and lack of vigor because of the effects of acid rain. Clearly, it is time to take action.

The strength of an acid is measured on a pH scale, which monitors the percentage of hydrogen ions contained in the solution. A perfectly neutral solution has a pH of 7. It is important to know that a solution with a pH of 6 is ten times more acidic than the neutral solution. A solution with a pH of 5 has a hundred times more acid. And so on. Thus when the pH of a lake drops from 7 to 4, there is a thousand times more acid in the lake than was originally present.

Well over 95 percent of the agents that put sulfur dioxide and nitrogen oxides into the air can be traced to four sources: coal-burning utilities, industrial boilers and heating facilities, nonferrous smelters, and gasoline- or diesel-burning vehicles. The United States and Canada together release 22.2 million metric tons of nitrogen oxides and 31.7 million metric tons of sulfur dioxide into the atmosphere each year (1.0 metric ton or tonne equals 1.1 short tons, or 2,000 pounds in the U.S. system).

While it is difficult to comprehend what a ton of smoke might look like—and while it sounds like a contradiction of terms—try to comprehend what 54 million tons of acid-producing fumes and smoke consists of and you have some idea of the scale of the problem. In the early 1990s the INCO smelter near Sudbury, Ontario, released 660,000 metric tons of sulfur dioxide into the air each year, and Noranda Mines in Quebec added another 538,000 metric tons. Both mine-related operations are located on the Canadian Shield, the region of Canada most susceptible to acid rain, and all of this SO_2 was transformed into acid precipitation.

One of the effects of acid rain on lakes is that fish fail to reproduce. Different fish species have different tolerances for acidic lake water. Lake trout is one of the most sensitive species, and reduced survival of the eggs and fry begin to appear when the pH of the lake falls below 6. Studies in southern Quebec show that nearly 75 percent of the fish species are extirpated as the pH of the lake approaches 5. Species of minnows and other small fish are frequently more sensitive than many sports fish. When the pH of a northern lake drops to 4.5, the lake will soon be fishless, and many other desirable life forms in the lake will soon disappear as well. Invertebrates most sensitive to acidic conditions include mayflies, caddisflies, stoneflies, amphipods, crayfish, snails, clams, and leeches—a number of which are important food sources for the predatory fish, birds, and mammals that live in our northern lakes and rivers.

The forests and lakes of Canada's national parks are far from immune from the acid rain problem. Acid rain crosses national park boundaries effortlessly. The problem is particularly serious for national parks in eastern Canada because they are downwind of some of the major industrial cities of Canada and the United States. Because natural rain reacts with CO_2 in the air, normal rain is always slightly acidic, with a pH of 5.6. However, the pH of rainfall in the national parks of eastern Canada has averaged between 4.3 and 4.8 for the past three years. All the national parks in or close to the taiga region of eastern Canada show this acidic precipitation, including Gros Morne and Terra Nova in Newfoundland, Cape Breton Highlands of Nova Scotia, La Mauricie and the Mingan Archipelago of Quebec, and Pukaskwa of Ontario. If the rainfall stays this acidic or worsens, major impacts on the lakes and forests of these and other parks can be expected.

The measurements are long-term averages for the pH of rainfall in these parks. Individual events, such as acid fog, can be significantly more acidic. Some U.S. national parks have measured acid fogs with a pH of 2.2. Another problem is that acid pulses may occur on lakes during the spring. Acidic snow builds up on the surface of a lake during winter, and runoff from melting snow in the uplands can be acidic. As a result, a sudden pulse of acid sometimes enters the lake when breakup occurs. Such strong acid pulses can devastate fish or amphibian eggs and cause major die-offs in minnow and invertebrate populations. A week later, this acid pulse has been neutralized by the natural limestone-buffering system found in many lakes, and no trace of the acid pulse can be found.

Lakes and forest soils have a natural buffering system to neutralize acid precipitation according to the amount of calcium carbonate and limestone that naturally occurs in the soils and lake beds of the region. Thus lakes located on limestone will be able to neutralize acid precipitation for a longer time than lakes located on the granitic bedrock of the Canadian Shield. Nevertheless, as acid precipitation continues, these natural buffering systems become exhausted, and lakes and forest soils reach a point at which their pH drops precipitously.

Once a lake deteriorates to a pH of 4.5 or lower, it takes on a strange sterilized look—devoid of fish and with very little plant growth. The water remains fairly clear, but leaves, silt, and muck build up on its bottom, and there is an absence of plants or animals. Lakes in this condition have been killed by acid rain; their aquatic ecosystems have been destroyed; and the normal healthy ecological processes of photosynthesis, grazing, predation, and decomposition no longer occur. The long-term future of these acid-killed lakes is unclear at this point, and research produces conflicting predictions. Some researchers believe that these lakes will remain in this state for years to come. Other researchers worry that acid-tolerant organisms—some of which could cause disease or be harmful parasites—will become established in these lakes. Much remains to be learned about the long-term future of lakes affected by acid precipitation.

Nor do the problems end there. Researchers document that, as a northern lake becomes acidic, mercury, aluminum, and other harmful metals are leached from the sediment and rock formations that form the lake basin. Thus as a lake acidifies, the mercury content in its fish increases significantly. A similar mechanism is at work in our forest soils. Researchers have documented that, throughout much of southern Quebec, sugar maple trees are stressed, and many are dying, from the effects of acid rain, which leaches aluminum from rock particles in the soil. In essence, Quebec's sugar maple trees are dying from aluminum poisoning, and it is a threat not only to the maple sugar industry but also to the landscapes of southern Quebec and eastern Ontario as we now know them.

Acid precipitation is also believed to cause respiratory health problems for people in eastern cities. Furthermore, the corrosive effect of acid precipitation on limestone and concrete buildings, bridges, and other facilities is believed to have caused more than a billion dollars worth of damage each year to these man-made structures. Scientists believe that we have not yet found all the deleterious effects of acid precipitation, and new problems will probably emerge with continued research.

The answer is not to try to neutralize the acid with limestone, covering our forest soils with it and pumping tons of limestone into our lakes each year. This is a Band-aid solution. The answer is to control SO_2 and NO_x emissions at their sources, and technologies exist that can do so. For electric-generating stations and smelters that use coal and other fossil fuels, a technique called flue-gas desulfurization, or FGD, has been developed. In FGD, wet limestone is sprayed into the plant's hot exhaust gases, where it combines with 90 percent of the sulfur dioxide before being emitted from the stack. The FGD technique is best used by generating stations and smelters already in existence—a retrofit solution. A more efficient technology for new facilities is called clean-coal technology. Clean-coal technology approaches vary. In one approach, the coal is pelletized and chemically treated to remove 96 percent of the sulfur dioxide before it is burned.

Vehicles must be changed to burn cleaner fuels. Alcohols such as methanol or ethanol are promising alternative fuels, as are gases such as propane or methane. It is striking that even after at least twenty years of intense effort to solve the acid rain problem, automobiles equipped for using propane or methane as fuels are still not commercially available in the North American market. Car owners must pay close to $1,000 to convert new vehicles to use propane or methane. The auto industry and petroleum companies have failed to show leadership on these badly needed initiatives.

On the other hand, there is some encouraging news. The United States and Canada have agreed on an international accord to reduce the output of nitrogen oxides and sulfur dioxide by 33 percent early in this century. Certain European countries have formed the Thirty Percent Club, agreeing to reduce sulfur dioxide emissions by that amount over the next several years. In some areas, the results are favorable. The pH of certain northern lakes is actually increasing back toward 7. Perhaps the corner has been turned. Perhaps the 1990s will be remembered as the decade when acid precipitation was at its worst.

Acid rain is a major environmental concern. Ten years ago it attracted wide media and political attention, but much of this public attention has died away. However, the problem of acid rain is still very much with us. It eats away at the buildings, bridges, and highways of our cities; it affects the trees growing in our backyards; it threatens the lakes and forests that surround our family cottages, and it threatens a vast sweep of the taiga in eastern and northern Canada. Acid rain affects all of us.

Maintaining the ecological health of our boreal lakes and rivers presents some daunting challenges. Much is at stake if we are to preserve the integrity and beauty of the 2 million boreal lakes that encircle the globe. Our success in this challenge can be measured in many ways. An important indicator of success will be how confidently our great-great-grandchildren will reach over the side of a canoe or boat, fill a cup, and drink refreshingly from these northern waters. That simple act will tell us a great deal about how we have treated our northern lakes and rivers.

General References

Irwin, R., ed. 1981. *Still Waters: The Chilling Reality of Acid Rain*. Ottawa: Minister of Supply and Services Canada, Subcommittee on Acid Rain.

Knystautas, A. 1987. *The Natural History of the USSR*. New York: McGraw-Hill.

Luoma, J. R. 1988. Acid Murder No Longer a Mystery. *Audubon* 90:126–35.

Mackay, W. C., ed. 1989. *Northern Lakes and Rivers*. Occasional Publication 22. Edmonton: University of Alberta, Boreal Institute for Northern Studies.

St. George, G. 1969. *Siberia: The New Frontier*. New York: D. McKay.

Schindler, D. W. 1998. A Dim Future for Boreal Waters and Landscapes: Cumulative Effects of Climatic Warming, Stratospheric Ozone Depletion, Acid Precipitation, and Other Human Activities. *BioScience* 48:157–64.

Struzik, E. 1992. The Rise and Fall of Wood Buffalo National Park. *Borealis* 3:10–25.

Young, S. B. 1994. *To the Arctic: An Introduction to the Far Northern World*. New York: John Wiley and Sons.

11

The Conservation of the Uncommon Loon

For many people the call of the common loon expresses the essence of the northern woods. It doesn't matter whether you are several days by canoe from the nearest road, camping on a remote lake of the Canadian Shield, or sitting comfortably inside a screened porch at the family's summer cottage. Suddenly a single loon overpowers the night with a drawn-out, haunting wail. Then a second loon answers at a distance, and then a third and a fourth. Yodels and laughter follow, rolling across the lakes and the gentle kame and kettle terrain of the taiga. Soon this cacophony of laughter and yodels melts into howling wails, and then all is silent again.

Such an outburst of loons on a quiet summer's evening stops most people in their tracks. The wild canticle has inspired generations of people because it expresses so well the beauty of the taiga. The Cree call the loon *mookwa,* which means *the spirit of northern waters*. Few people who have stood next to a calm northern lake at night and listened to a chorus of loons calling one to another will question the appropriateness of this name.

Summer in the taiga without the call of loons is a dismal proposition. And yet this siren song of the taiga is in trouble. The common loon is not faring well in the contemporary world, and while its problems are significant, they are often subtle and overlooked. Many people who cherish having loons on the lake by their summer cottage are unaware that "their" loons have not raised offspring in years. The great northern diver, as this bird is known in Britain, often suffers from its interactions with humans. Thus its problems are most pronounced along the southern border of its range, where human populations are the densest. To understand these problems and what can be done to solve them, we need to understand the evolution and ecology of the common loon.

Primitive Bird or Specialized Hunter?

Paleornithologists believe that loons are closer to birds' ancestral reptilian stock than almost any other living avian species. They are large, heavy-bodied birds; their

The common loon, a large, heavy-bodied bird, is marvelously adapted for diving.

skeletons contain few air sacs and are mostly made of solid bone. Loons become airborne only with great difficulty—by running across the surface of the lake. Their flight is direct and fast, and they are unable to maneuver in flight as adroitly as many bird species. For these and other reasons, some biologists believe that loons have changed little over millions of years, and thus they view loons as the most primitive of North American birds. However, when one looks closely at some of the loon's diving abilities and hunting skills, one begins to see not so much a primitive bird as a highly specialized aquatic predator.

In North America, there are five species of loons, namely, the yellow-billed loon *(Gavia adamsii)*, Pacific loon *(Gavia pacifica)*, Arctic loon *(Gavia arctica)*, red-throated loon *(Gavia stellata)*, and common loon *(Gavia immer)*. The first three species nest mostly on medium-sized and large lakes on the tundra. The red-throated loon, the only loon capable of taking flight from land rather than running across water, often nests on small tundra ponds. These ponds freeze to the bottom each winter, but because red-throated loons can take off from land, they often fly to nearby larger lakes or to the ocean to forage. Like other loons, the red-throated loon lands breast first on the surface of the water, not on ground.

The fifth species, the common loon, is much more far ranging than the others, nesting not only throughout the southern tundra but also throughout the entire taiga region of North America. The breeding range of the common loon, similar to the range of the moose, is nearly synchronous with the distribution of the boreal forest in North America. Both are fitting flagship wildlife species for the taiga biome.

Common loons migrate in the fall, spending the winter feathered in a drab gray-and-white plumage on the Great Lakes or along a considerable stretch of coastline, ranging from the Aleutian Islands to Baja California and from the coast of Newfoundland to the Gulf of Mexico. Juvenile common loons may spend two or three years in these areas before they migrate northward. These adult and juvenile loons feed on individual feeding territories offshore during the day, but they gather into large flotillas of loons at night, reflecting their underlying social nature.

All five species of loon are elegantly adapted for diving and catching fish. In the air, buoyant and maneuverable they are not, but loons are virtual killer submarines on or under the water. Many of the features that some biologists interpret as primitive can equally be understood as adaptations for their aquatic, predatory existence. Loons are much larger than most people realize. They are the size of a bald eagle, and they weigh as much as a large Canada goose. Adult common loons weigh on average eleven pounds (5 kg).

A common loon is a heavy-bodied, torpedo-shaped bird that propels itself underwater with great speed, using webbed feet set far back on its rump and powerful leg muscles—all attributes that make the loon a powerful swimmer and excellent underwater hunter. On land, those same legs, because they are set so far back on the body, are nearly useless, so that to move across land an adult common loon must either hump along on its breast using its folded wings as crutches or shuffle along slowly in an upright, waddling gait. The position of its legs also dictates that most loons cannot land on ground at all nor in the water feet first, like a duck. To land, the common loon touches down on the surface of a lake breast first, slicing into the water with incredible grace.

The bones of a loon are solid, lacking the elaborate air-sac system of most birds. This may be a primitive, ancestral feature, but it also increases body density and allows the loon to dive to great depths. Common loons in the Great Lakes have been recorded at depths of 600 feet (183 m). Humans diving to these depths need a specialized diving bell, need to breathe a special mixture of oxygen and helium, and must spend many hours in a decompression chamber upon returning to the surface. Common loons have been observed to make dives like this several times during the same hour.

Environmental Threats

Common loons can squeeze much of the air out of their feathers and partially deflate the few air sacs that they do possess, thus "trimming" themselves to float at any level on the surface of the water. While canoeing, I have had the experience of being watched by a nearby loon that is completely submerged except for its head. When I came too close, the loon lowered its periscope and disappeared like a submarine, leaving hardly a ripple on the surface of the lake.

Common loons show other sophisticated diving adaptations. They are able to regulate their oxygen-rich blood, reserving most of it for their central nervous system while their muscles show an unusually high tolerance for the accumulation of carbon dioxide—all important physiological attributes for long and deep underwater dives. By comparison, gulls and terns are elegant fliers, but they are so light bodied that they are incapable of diving under the surface of the water. On the other hand,

Ring-billed gulls are light-bodied, elegant flyers but are incapable of diving under the surface of the water.

the aquatic adaptations of the loon carry constraints for flight. Their heavy bodies become airborne only with great difficulty and are capable of only straight, direct, and rapid flight. Such seems to be the environmental costs of being a diver.

Because the common loon is unable to take off from land, it must run athletically over the water for a considerable distance, wings flailing, to get enough lift for takeoff. These characteristics led, at the beginning of this century, to a wasteful type of sports hunting. A. C. Bent, the famous ornithologist, writing about 1920, stated: "It is a constant temptation to all gunners to shoot at passing loons, for they are swift, strong flyers and are very hard to stop." Also in the early part of the century, before the International Migratory Bird Treaty of 1912 prohibited it, there was another type of unfortunate loon hunting, called pond shooting. Allan Pistorius describes this barbaric sport:

> The glory of pond shooting depended on the belief that the loon, upon seeing the flash of the rifle, could dive before the bullet could reach it. Surrounded by hunters, and yet unwilling to risk the labored run necessary for flight, a loon in this situation dives and dives and dives until exhaustion slows the reaction and a bullet ends the game.

The hunting of loons is thankfully a thing of the past: There is no legal hunting of loons in the United States or Canada. Harassment of loons by fishers is also largely a thing of the past. Some anglers have mistakenly believed that loons harmed sports fishing by preying on game fish, but research documents that availability is the key to the diet of the common loon. Thus these loons more frequently prey on fish that are slow moving and hence easier to capture, "rough fish" rather than game fish. Loons also feed on crayfish, fresh-water shrimp, snails, frogs, aquatic insects, and some aquatic vegetation.

Many of the loons' contemporary problems materialize during winter, when the birds occupy feeding territories on the Great Lakes and along both coastlines. Here they feed, preen, and doze during the day and form large flotillas at night. These offshore winter feeding activities cause some loons to become entangled in commercial fishing nets. Loons are also exposed to toxic chemicals from ships and industrial effluents and to diseases, such as botulism.

Oil slicks—whether they are from massive accidents such as the Exxon Valdez or intentional bilge-cleaning operations from sea-going ships—increase the winter death count of loons. Technological processes sometimes take a staggering toll. For example, in the waters of the Gulf of Mexico off the western coast of Florida, as many as 7,000 common loons died during the winter of 1982–83, presumably from complications following mercury poisoning. Epidemics of botulism caused by food contaminated with the highly toxic chemical produced by the bacteria *Clostridium botulinum* have killed large numbers of loons. Eight epidemics of botulism, killing thousands of loons, occurred on Lake Michigan between 1960 and 1985. More recently, since 1998 hundreds of loons have been found dead on the shores of Lake Erie. It is believed the loons are dying after eating gobies, small fish introduced into the lake from the ballast water of Eastern European ships. Gobies eat another nonnative species, the zebra mussel, which researchers think is concentrating botulism spores that ultimately infect the loons. Large numbers of mergansers, cormorants, and gulls have been dying as well.

These environmental threats make winter the season of highest mortality for adult common loons and other water birds and steadily reduce their breeding populations across North America even in areas as remote as interior Alaska.

Loyalty to a Lake and a Mate

In the spring, most loons migrate northward and inland, returning to the same lake each year to breed—and to face new challenges. The reproductive biology of the common loon is central to understanding its conservation. A pair of common loons is believed to mate for life, and these birds will return to the same lake year after year until death ends the union.

In avian societies, the loon's mating pattern is the rule rather than the exception. Monogamy (meaning that one male mates with one female and both assist in the raising of the young) is observed in approximately 90 percent of all bird species. But the degree of monogamy varies. The exclusive pair bond may last for a single nesting (such as in the house wren), an entire breeding season (most passerines or perching birds), several successive breeding seasons (some pairs of American robins and tree swallows), or life (swans, geese, eagles, and the common loon). Common loons thus exhibit a high degree of loyalty to their mates and breeding territory. Interestingly, however, the pair does not migrate together, nor does it seem to spend the winter together. These patterns are partially explained by the fact that one of the pair usually leaves the breeding territory during early to mid-September while the other parent stays on the territory until the chicks are ready to migrate—usually more than a month later. The departure of one parent a month early is probably an adaptive behavior to maximize the food available to the chicks before their southward migration.

The following spring, the true strength of the pair bond expresses itself. Each member of the pair migrates northward and returns to the same breeding territory; courtship commences, and the pair works together to raise one or two chicks on their lake that summer. Loons in the wild can live to be over twenty-five years of age, and the same pair of loons has been observed to raise chicks on the same territory for up to twenty years. If one mate dies, the pattern normally observed is that the other bird returns to the same territory, does not mate with another bird, and patiently waits summer after summer for the other mate to return.

Variation is the raw material for natural selection. Variation in innate behavior lets natural selection test which behaviors help to leave more offspring and leads to the evolution of new adaptations. It would be surprising to see the common loon practice one mating system exclusively without variation, and recent research suggests that such is not the case. A study in Isle Royale National Park, in Lake Superior, involving banded common loons, documents that approximately one-fifth of the birds took on a new mate or changed territories once one of the mates had died. If these behaviors are at least partially inherited and result in more offspring, we can expect these uncommon behaviors to become more prevalent among loons in the future.

Common loons begin to breed relatively late for an avian species. During its first autumn a young loon migrates with one of its parents to that parent's wintering ground, normally the ocean coasts or regions of the Great Lakes that seldom freeze. The juvenile loon usually stays in this area until just before it is three years old, then it migrates inland and tries to find a vacant breeding territory. A pair of loons raises a maximum of two young each year and, even under natural conditions, do not raise young every year. If either the breeding territory or their physiological status is inadequate, common loons often forgo nesting, typically nesting three of every four years. Clearly, common loons even in the best of times do not produce a surplus population.

On two occasions I had the privilege of observing the springtime reunion of a pair of loons upon their return to their breeding territory—a small lake in Prince Albert National Park, in northern Saskatchewan. The reunions occurred shortly after breakup, when a portion of the lake remained covered by ice. First, one loon arrived on the lake, and a day or two later its mate arrived. Soon after the arrival of the second mate, both birds displayed to each other in a manner that can only be described as ebullient. There was much splashing and rushing toward each other. There was jumping but also quiet, mutual bill dipping, as they swam side by side. There was their strange "penguin dance," in which they rise up vertically on the surface of the water, treading water and circling each other or facing each other with wings fully extended and their bills pressed tightly against their chests. This intense displaying continued for several days. The return to the family territory and the reunion with its mate are obviously special moments in the life of a loon.

Sizing up the Family Territory

Although common loons occasionally nest on kettle lakes of less than twenty acres (8 ha) in size, they usually prefer much larger bodies of water. A lake of a hundred acres (40 ha) or more is typically claimed by a single pair of breeding loons.

Loyalty to the breeding territory is also the Achilles' heel for common loons when it comes to interactions with humans. If a pair of loons returns to its lake to find a cottage, fishing camp, or campground being constructed too near its breeding territory, the pair usually continues to summer on that lake but ceases to breed. People living on the lake continue to enjoy loon music during summer evenings without realizing that the reproduction of these magnificent birds has failed. Hundreds of recreational lakes throughout the northern states and southern Canada are inhabited by loon pairs, yet many of these loons never raise young—they are loon families destined for extinction. And it is the more tragic because the whole matter might be resolved by a few simple, thoughtful actions on the part of the people who use the lake.

The critical sign that people should look for is not whether adult loons appear on the lake year after year but whether young loons are seen each summer. They are not difficult to spot. Beginning in early July, they can be observed riding on a parent's back. Later in the summer, they can be seen swimming close to one of their parents while the other parent feeds or patrols in a more distant part of the territory. If young loons are not observed by early August, it is valid to conclude that some factor has sabotaged the reproductive efforts of the mated pair.

The causes can be any of a number of factors and often affect the chicks more than the adult loons. Robert Alvo, who studied loon reproduction on eighty-four lakes in Ontario, found that 62 percent of the loon broods died on acidified lakes, compared to 14 percent on healthy lakes. Although adult loons can fly to neighboring lakes to feed themselves, their young cannot. Alvo concluded that young loons raised on lakes affected by acid rain often simply starve to death.

More is at stake than just the loss of a few loons to the loon population. Because loons are so susceptible to disturbances in their environment, they are important indicators of the ecological health of their lake. A lake that produces one or two young loons for each mated pair can probably be given a good bill of health. If the loons raise their offspring successfully despite the threat of acid rain, cottages, sewage systems, and recreational activities, it is fairly safe to assume that animal species more adapted to human impacts are probably doing equally well. Healthy loons usually mean a healthy lake. Because this is the case, let's look at the breeding requirements of the common loon in closer detail.

Living with Loons

As mentioned, adult loons arrive at their lake shortly after the ice goes out, and courtship and breeding usually last a week. Loons court in the water, but they return to the land to breed. Water will support ducks in the act of copulation but not the heavy-bodied loon. There's a touch of irony that the common loon, an avian species almost perfectly adapted for an aquatic existence, must seek out land in order to breed.

Loons' nests vary from simple scrapes on bare ground to huge flat structures made of aquatic vegetation. Islands are favored, but shoreline nesting sites are also used. Loons have even been known to nest on top of old muskrat houses or on floating mats of vegetation. Invariably, the nest is close to the shoreline near water that is at least three feet (1 m) deep—a depth that allows the loon a quick underwater escape if necessary. The female lays two eggs, which are brownish-olive in color, with dark splotches. The eggs are huge, larger than the eggs of birds twice the size of a common loon, and the laying of the eggs is protracted and appears painful for the female. The parents take turns in incubating the eggs—twenty-eight or twenty-nine days—rotating turns in two- or three-hour shifts. The parents not only warm the eggs but also protect them from predators, such as crows, ravens, gulls, mink, otter, skunks, and muskrats. More recently another predator—the raccoon—has become an effective predator of loon eggs. The extension of the range of the raccoon northward has devastated the breeding success of the common loon along the southern boundary of its range. Raccoons thrive in cottage country, often at the loon's expense.

Humans can negatively impact the breeding success of loons in a number of ways, but these adverse effects can largely be avoided by observing some loon etiquette rules during the spring and early summer. Ironically, canoeists or fishers—a group that is usually ecologically aware—are likely to have the most serious impacts on nesting loons. Let's examine what happens.

Loons nest close to the shoreline. Because of the low temperature of the lake water, this nesting site is a remarkably cool microenvironment. When an angler trolling the shoreline or a person paddling a canoe comes within several hundred yards of the nest, the incubating loon quietly slips off the nest and dives under the water, surfacing farther out in the lake, well away from the nesting site. It does not return to the nest until the intruder has moved a good distance away. Often these intruders do not even know that a loon's nest is nearby. If they are stopping to cast for an extended period in a promising small bay or canoeing out to a favorite island for an afternoon picnic, the time that the parent loon stays off the nest can be protracted. The eggs cool to a dangerously low temperature, and the embryos may never recover. Alternatively the eggs may be spotted and destroyed by ravens, gulls, or hawks flying overhead.

The loon incubating its eggs can also be surprised by a passing motorboat and may leave the nest in such alarm that one or both of the eggs are knocked into the water and thus permanently lost. Parent loons are not able to retrieve floating eggs and return them to the nest. The wake from motorboats passing close to shore have also been known to wash loon eggs into the water.

A Helping Hand

The human threat to loons can be minimized by human cooperation. Residents of many lakes have organized to protect the loons on their lake by posting vulnerable nesting areas with buoys and signs to keep people and boats away from them during the critical six-week period of incubation. If canoeists, fishers, and powerboaters stay away from these nesting sites between the time the ice leaves the lake until the chicks hatch and enter the water, the breeding success of loons can be markedly improved.

For lakes that have a serious problem with raccoons marauding loon nests, an artificial nesting platform can be constructed. Research shows that raccoons usually will not swim out to an island to steal eggs. During the 1980s the Loon Preservation Committee of New Hampshire spearheaded this research and had great success with these artificial islands. At present approximately 20 percent of the young loons successfully raised in New Hampshire are incubated and hatched on artificial platforms. A nesting platform is not difficult or expensive to build. It is made of four dry cedar logs, approximately six feet (2 m) in length, joined together to form a square. The square is covered with wire fencing, on top of which a chunk of natural shoreline vegetation, soil and all, is placed. The platforms are anchored in a sheltered cove with concrete blocks. Loons are often attracted to these artificial islands, and the passing years have shown that their nesting success can be quite high when they use them. Wildlife programs from Minnesota to New Hampshire use artificial nesting platforms as a way to encourage loons to nest in locations free from human disturbance and raccoon predation.

Once the two eggs hatch (they do so usually about a day apart, and the chicks enter the water shortly after their feathers dry), the loon family moves to a marshy nursery area, which may be some distance from the nest. The nursery area is an important part of the family territory. Ideally, it contains water that is clear and shallow enough to limit the size of predatory fish; is in an area of the lake reasonably free of eagles, osprey, and gulls hunting overhead; and is rich enough to furnish an eleven-week supply of food for the two growing chicks.

At this point in their development, the tiny chicks, looking like dust balls with beaks, peck at vegetation, preen their downy feathers, chirp like newborn chicks, and eat the minnows and tiny fish that their parents bring them. Although young loons are capable of short dives, they are vulnerable during the first few weeks of their life. During this period, they spend much time riding on their parents' backs, a strategy that prevents both heat loss and predation. If humans intrude into the nursery area, the chicks normally hide along the shoreline or flatten out along the surface of the water, while one or both adults leave them for the open water. Sometimes the adult loons hang back and call anxiously; other times they engage in dramatic surface rushes, trying to lure away the disturbers. After many intense tremolos and a desperate racing back and forth across the water, the adult birds may give up from exhaustion.

As young loons gain strength day by day, there is one last danger to be reckoned with: the powerboat crowd. The danger is sometimes premeditated—unfortunately, a small fraction of powerboaters enjoy chasing loons. Other powerboaters, however, may run down and kill loon chicks without even realizing it. Although adult loons can easily avoid the onrushing boats, loon chicks exhaust easily and are not strong swimmers. While loons will sometimes nest again after losing a clutch of eggs, they do not do so after losing chicks, and another year may slip away with no new loons on the lake.

The decline in loon populations along the southern portion of its range has been precipitous. In the northeastern United States loon populations were reduced to between 35 and 75 percent of their previous levels in twenty short years, from 1965 to 1985. New Hampshire's situation is well documented and typical. The loon population bottomed out in New Hampshire during the summer of 1978, when only eighty pairs of loons established nesting territories throughout the entire state. Fifty-seven of these pairs attempted to nest; only twenty-nine pairs successfully raised chicks, and only thirty-four of those chicks survived to autumn. Thus in a state with hundreds of lakes, only a handful of common loons were produced.

Since 1978 the North American Loon Fund and its affiliated groups have turned the situation around. New Hampshire's Loon Preservation Committee has organized volunteer groups of lake cottagers to carry out many of the suggestions described above. These volunteer citizen groups (some call themselves the Loon Rangers) monitor loon nesting sites on their lakes and warn off boaters or canoeists who do not heed the buoys or warning signs. Scott Sutcliffe, who helped organize these efforts, tells of receiving a 5:30 A.M. phone call from a nest watcher, who reported, "Scott, you're an uncle." When he got to the lake an hour later, the cottagers were on the dock drinking orange juice and champagne.

Loon Rangers and champagne orange juice may sound a bit corny, but they're important. Research shows that the actions of people such as these help maintain a loon population on lakes heavily used by humans. However, progress can be painfully slow. In New Hampshire, although the number of nesting pairs known to raise at least one chick in most years has more than doubled since the Loon Preservation Committee began its work in 1975, it represents over twenty-five years of dedicated work. Furthermore, other problems that adversely affect loons—acid rain, lead poisoning, and mercury contamination—are beyond the reach of small local groups. Nevertheless, progress can be made on all of these problems if there is the dedication and political will to do so.

I am reminded of an experience I had in northern Sweden over a decade ago. I was visiting Swedish colleagues, investigating environmental issues of common concern. One evening a local forester came by to report that a loon had been sighted on a nearby lake. The group became quite excited and decided then and there to set off for the two-hour drive to the lake to see the loon. For several of my Swedish friends it was the first loon they had seen in years. The experience haunts me. Given several more decades of development and economic expansion, will Canada follow in Sweden's tracks? Will our loons become as scarce as loons in Sweden? Is this the future of the spirit of northern waters? If not, what must we do to give mookwa a different fate?

General References

Alvo, R. 1986. Lost Loons of the Northern Lakes. *Natural History* 94:60–64.
Bent, A. C. 1919. *Life Histories of North American Diving Birds*. New York: Dover.
Dunning, J. 1985. *The Loon: Voice of the Wilderness*. Dublin, N.H.: Yankee Publishing.
Ehrlich, P. R., D. S. Dobkin, and D. Wheye. 1988. *The Birder's Handbook*. New York: Simon and Schuster.
Loon Preservation Committee. http//www.loon.org.

McIntyre, J. W. 1976. 1983. Nurseries: A Consideration of Habitat Requirements during Early Chick-Rearing Period in Common Loons. *Journal of Field Ornithology* 54:247–53.

———. 1988. *The Common Loon: Spirit of Northern Lakes*. Minneapolis: University of Minnesota Press.

North American Loon Fund. http://facstaff.uww.edu/wentzl/nalf/analfhomepage.html.

Pistorius, A. 1979. Feathering the Loon's Nest. *Country Journal* 6:42–46.

Sutcliffe, S. A. 1978. Changes in Status: Factors Affecting Common Loon Populations in New Hampshire. In *Transactions of the Northeast Section of the Wildlife Society, Thirty-fifth Northeast Fish and Wildlife Conference*.

12

The Land That God Gave Cain

It is time to change bearings and beat upwind. To this point, I have been exploring the natural history and ecological processes of the Canadian taiga. Now let's focus on the human beings who have to support themselves and their families there. How best is that done, given the harsh climate of the taiga and the restricted productivity of its lakes and forests? What would a regional economy of the boreal forest look like if it were compatible with its ecology? What are the alternatives? How do humans best live on this land without sabotaging its fragile ecology? In this chapter I argue that the present economy of the Canadian taiga is based on the exploitation of natural resources at levels that cannot be sustained and explore the underlying attitudes that foster this overharvesting of resources. In the following chapter I examine alternatives—that is, what a truly sustainable economy in the Canadian taiga might look like.

Consider the following statistics. During the 1990s Canada mined more nickel and zinc than any other country in the world. It is among the world's top five producers of asbestos, potash, sulfur, uranium, gypsum, gold, aluminum, copper, cadmium, lead, iron, silver, and platinum. In 1987, for example, Canada's mining industry produced over $16 billion worth of minerals (excluding oil, natural gas, stone, and gravel). Many of these mines are located in the boreal forest region of Canada, the majority on the Canadian Shield. By its very nature, mining depends upon nonrenewable resources, a mining industry lasts in a region only as long as there are economically viable minerals to exploit, refine, and market.

Canada is one of the most productive fishing countries of the world, both in tons of fish caught and tons of fish exported. Canada's success is based on its numerous bodies of water with their abundant fish. These include not only Canada's three oceans but also its more than a million boreal lakes and its many rivers. In 1985 Canadian commercial fishers harvested and sold approximately $115 million of freshwater fishes, largely taken from taiga lakes. Is the freshwater fishing industry in Canada operating on a sustainable basis? David Schindler, recognized as one of Canada's leading aquatic ecologists, pointed out in a recent review that virtually every lake accessible by road in Canada's taiga has been overharvested. Thousands of lakes accessible only by air have been overharvested as well. In addition, acid precipitation is rendering thousands of lakes in northern Ontario, Quebec, and the Maritime Provinces incapable of supporting any fish.

The forest industry is the second largest industry in Canada (only agriculture is larger). This industry includes logging companies, pulp and paper mills, sawmills, veneer mills, shingle mills, and such wood-using enterprises as plywood and chip board mills. In 1996 the forest industry in Canada's taiga region produced $55 billion worth of products, harvested nearly 21.5 million acres (868,000 ha) of forest, and provided 830,000 jobs. However, although approximately one of every five jobs held by Canadians is in the forest industries, these jobs are declining as the industry becomes more automated. In the boreal forest region from 1975 to 1996, the number of direct jobs in the forest industry decreased by 15 percent.

Is the harvesting of Canada's boreal forest done on a sustainable basis? Approximately 100,000 square miles (260,000 km^2) of boreal forest (an area larger than all of Great Britain) have been clear-cut, and only a meager 3 percent of it has been replanted by hand, leaving the rest to natural regeneration and the vagaries of erosion. The Sub-Committee on the Boreal Forest of the Canadian Senate in a 1999 report summarized the situation as follows:

> The demands and expectations placed on Canada's boreal forest have escalated to the point where they cannot all be met under the current management regime. Highly mechanized timber harvesting is proceeding at a rapid pace, as is mineral and petroleum exploration and extraction. At the same time, the boreal forest is being asked to provide a home and way of life for aboriginal communities, habitat for wildlife, an attraction for tourism and a place where biodiveristy and watersheds are protected....
>
> In the face of these demands, expectations and threats, Canadians must come up with new and better ways to manage our activity in the boreal forest to meet the competing realities of preserving the resource, maintaining the lifestyle and values of boreal communities, extracting economic wealth, and preserving ecological values.

Different Perceptions of the Taiga

It is tempting to attribute this excessive harvesting of the natural resources of the Canadian taiga to a well-developed capitalistic spirit or to the dominant Judeo-Christian culture steeped in the biblical maxim: "Be fruitful and multiply, and fill the earth and subdue it; and have dominion over the fish of the sea and over the birds of the air and over every living thing that moves upon the earth" (Genesis 1:28). Both capitalism and some of our traditional beliefs have undoubtedly contributed to the exploitation of the taiga, but I believe that there are other attitudes that have directly contributed to it—and that need to be challenged. There seems to be a deep-seated, uncritical attitude among North Americans in general, but particularly Canadians, to look upon the taiga as an impoverished land, a remote and nearly worthless part of the continent. Too many Americans and Canadians perceive the boreal forest as "just the bush," "an economic wasteland," and "a land of little sticks."

For three hundred years Canadians as a society have supported the principle that if a person or a company can figure out how to make a profit off this land, they should be allowed to do so, and more times than not the federal and provincial governments have been willing to subsidize or be co-owners of the enterprises. Mines extracting everything from aluminum to zinc, hydroelectric dams such as the massive James Bay project in Quebec and the W. A. C. Bennett Dam in British Columbia, provincially owned paper mills, freshwater fishery marketing boards have all harvested the taiga at unsustainable rates—and all with the apparent blessing and political support of the majority of Canadians.

This view of the taiga as an economically barren land has deep, historic roots. Samuel de Champlain, the explorer and founder of New France, wrote back to his king describing the boreal forest of Quebec as "the land that God gave Cain." In 1670, when King Charles II of England established the Hudson's Bay Company, he gave this "company of adventurers" complete control of approximately 3 million square miles (7.7 million km^2) of Rupert's Land, which included all the land drained by rivers flowing into Hudson Bay. The king asked for no taxes and no royalties; he simply gave away this immense territory to the Hudson's Bay Company as if he was glad to be rid of such a wretched land.

As I pointed out in the Preface, not much has changed during the past three centuries. When the government of Alberta in the late 1980s signed the forest management agreements, mostly with foreign-owned corporations, they gave away most of the forests of northern Alberta, asking for little else than that jobs might be created. Any royalty payments were years off, and in fact the tax breaks and subsidies to these large companies cost the taxpayers of Alberta millions of dollars for the privilege of letting Japanese corporations clear-cut the forests of northern Alberta. Similar giveaways have occurred in British Columbia, Saskatchewan, Manitoba, and Ontario. Likewise, Quebec and Newfoundland and Labrador have similar histories of giveaways of boreal forest resources.

It doesn't have to be this way. As noted in the Chapter 1, the taiga is remarkably similar in its circumpolar sweep. There are differences, and the species do vary region to region, but the ambience of the forest is the same. For instance, in all parts of the taiga similar animals and plants occupy similar habitats and niches, even though species may vary region to region. To mention a few examples, there is normally a crow-sized woodpecker that drills for carpenter ants in the center of softwood coniferous trees, there are always insectivorous plants growing in taiga bogs, and there is always a rich community of wood warblers and other small birds harvesting the diverse and dense insect populations of the taiga. In addition, the taiga's defining characteristics, such as its highly simplified species composition, its abundance of peatlands, and its predominance of coniferous trees are consistent right around the globe.

As a result of this ecological affinity, we might expect that northern peoples would perceive the taiga in similar ways and that societies would manage its resources in similar manners. My exploration of this topic over the past decade finds that this is definitely not the case. The Canadian and American perceptions of the boreal forest are different from the Russian perception of the taiga, which differs again from the Scandinavian and Finnish views. Aboriginal peoples throughout the circumpolar taiga have perceptions of their homeland that differ yet again. Let's explore three of these worldviews of the taiga.

The Aboriginal Perception

Perhaps the antithesis of viewing the boreal forest as a cursed and worthless land is found among the aboriginal cultures of northern Canada and Alaska, whose perception of the forests, lakes, and rivers of the taiga is radically different. One of the best in-depth descriptions of northern natives' perception of their boreal environment is in Richard Nelson's book, *Make Prayers to the Raven,* written after the author had lived and worked among the Koyukon people of north-central Alaska for several years. Nelson uses his skills as an anthropologist to describe the Koyukon's annual round of activities and their cultural perception of the taiga. In doing so, he captures the worldview of all the Athabascan people, who occupy the boreal forest from northern Alaska to northern Saskatchewan.

Nelson poses a central question: How is it that the Koyukon have lived on their land for thousands of years and yet have altered it so little? Fossils and the pollen record suggest that this land has changed little during the past 8,000 to 10,000 years of human habitation. Nelson ultimately attributes this to the harmony that exists between these people and their land, a harmony based on their spiritual beliefs and cultural perceptions. Many of these relationships are determined by the fundamental Koyukon belief that the natural and supernatural worlds are inseparable, each being intrinsically a part of the other. Everything in nature, they believe, is conscious and watchful and imbued with spiritual powers. Humans and natural entities are involved in a constant spiritual exchange, and this interchange profoundly influences and determines how humans act toward the natural world.

In the traditional Koyukon understanding of their boreal homeland, there is only one society, and all plants, animals, physical objects of the environment, and human beings are members of that society. Everything that humans take from nature—be it salmon, moose, building materials, or wood to heat their homes—is received as a gift, provided by the spiritual forces that inhabit the natural world. Hundreds of spiritually based rules govern all human interactions with the spiritual beings that make up the Koyukon's boreal world. The purpose of these rules is to show respect and to avoid disrespect. It is an extremely complex system of behavior, and this code of conduct is passed on from elders and parents to their children and grandchildren. Nelson states: "Proper human treatment of the natural world is enforced by the omnipresent threat of retribution by potent spiritual powers. This retribution can impair an offender's ability to harvest resources or it can bring illness, and possibly death, to the offender or a near relative." Perhaps only by adopting a philosophical paradigm that is more compatible with this aboriginal worldview will our society come to live in sustainable ways with our environment. Gaia and Deep Ecology are examples of contemporary ecological movements that are exploring ways to accomplish this important change in perception.

The Scandinavian and Finnish Perception

Norway, Sweden, and Finland also contain large tracts of boreal forest, and the taiga in each of these societies is part of its identity and culture. I focus mostly on Sweden because I know it the best, but equally interesting examples could be given from Norway and Finland.

Two-thirds of Sweden is forested, and over half of the country is boreal forest. Swedish families are accustomed to regularly spending time in a natural setting, either picnicking in the woods, or walking along a lakeshore, or cross-country skiing in the mountains. Nature is such an important part of Swedish life, and Swedish culture has always integrated the natural world with traditional practices and even religious beliefs.

One expression of this marriage of natural phenomena with religious celebrations is Santa Lucia's Day, one week before the winter solstice, when young girls all over the country wake up their family on that dark winter morning dressed in white and wearing a wreath in their hair decorated with lighted candles to mark the return of the Sun. Another example involves the European brown bear, which Swedish folklore recognizes as an intelligent animal with many human characteristics. Upon encountering a bear at close range, tradition holds, one should say, "Bear, you are not baptized from the same font as I am." This is a polite way of telling the bear that you know that it is unbaptized, whereupon the bear, suffering severe embarrassment, runs off into the woods, leaving you out of danger and free to go on your way.

A Swedish folk belief about the Arctic loon (*storlom,* in Swedish) is especially touching. As recently as two generations ago, there was in Sweden (as in most Western societies) a tremendous social stigma against a woman giving birth to a child

Bears, such as this American black bear, traditionally have been viewed as powerful and intelligent animals.

out of wedlock. Unwed women, upon giving birth, might even drown their babies, who, being unbaptized, would be cursed to wander the Earth instead of entering Heaven. On calm summer nights, the mournful calls of the storlom echoing across hills and quiet lakes were believed to be the cries of these dead babies, reminding their mothers of their terrible deeds.

In contemporary society, cultural affinity for the boreal forest gets expressed in contemporary ways. For instance, a popular Swedish saying has it that in the heart of every Swede is a small forest cottage, painted barn-red with white trim. When asked about this adage, Swedes typically laugh lightheartedly, but then they agree, stressing that the cottage is an important part of Swedish life. The cottage is where people renew their relationship to the land. Oftentimes the forest cottage has been in the family for several generations, and a unique sense of family history permeates the premises. Families often gather there; it is where family members pause to witness and share the passing of the seasons, if not the years. As mentioned in the Preface, the *dacha, mókki,* cabin, or forest cottage—whatever name it goes by—is an important part of many northern cultures, and it plays an immensely important role in the life of many contemporary Swedish families.

The Swedish perception of nature, and therefore of the taiga, is thus different from the typical Canadian perception of it as a remote and largely worthless region. In fact, Swedes commonly refer to the taiga as the Swedish national forest and view it as their natural heritage and a foundation of the Swedish culture. But despite their appreciation for the taiga and its northern ecology, Swedish management of their boreal forest is not without major problems or mistakes.

Some of the largest clear-cuts in the world are in northern Sweden: I have stood in clear-cuts that are ten miles (16 km) across. Swedes have carried out so-called silvaculture experiments that have left large tracts of productive lands occupied by only stunted, twisted trees. The use of herbicides and the fertilization of forested lands using urea compounds have devastated the reindeer industry and caused economic hardship among the Sami, not only in Lapland but also throughout northern Sweden and Finland. Deep tillage, a clear-cut reclamation process that is practiced in many areas of northern Sweden, involves ploughing a cut-over area in furrows, much like a cultivated field. At its most intense use, deep tillage leaves furrows that are nine feet (3 m) in depth, and the erosion and disruption of the soil is considerable. Due to public pressure and poor results, Swedish forestry companies have modified this practice and now are ploughing furrows only a foot (25 cm) deep, but the practice is still damaging.

The drainage and fertilization of wetlands carried out in both Sweden and Finland is another ruinous forestry practice. Drainage ditches often six feet (2 m) deep are dug through and around a wetland in order to drop the water table. The area is then planted with fast-growing, often nonnative conifers, and the area is repeatedly fertilized. This program has decimated thousands of peatlands in Finland and Sweden, all in an effort to grow trees faster in order to supply these countries' giant pulp and paper mills. Much research has been carried out—particularly in Finland—on the environmental effects of bog drainage. The research consistently finds in these areas a worsening of water quality and reduced populations of fish, birds, mammals, and plants. Furthermore, significant amounts of carbon dioxide are released into the atmosphere from the decomposition of peat; sphagnum peat is 90 percent carbon, and as it decomposes, CO_2 is released, contributing to the greenhouse effect and global climate warming.

It is important to remember that bogs and other peatlands act as sinks for carbon; some bogs have been accumulating carbon in their peat deposits since the glaciers receded. A quarter of the world's pool of carbon contained in soils is stored in boreal and wetland ecosystems. If all of the world's peatlands were oxidized, approximately 336 gigatons (370 gigatonnes) of carbon would be released into the atmosphere, causing major disruption to global weather systems and large-scale extinction of plants and animals in all terrestrial ecosystems. Draining bogs and other peatlands to grow more trees for an unsustainable forest industry is like burning the furniture to keep warm. In a recent critique of these practices, Bill Pruitt concluded: "It is clear that bog drainage must not only stop but the myriad present drainage ditches must be plugged and the bogs allowed to resume their vital role as carbon sinks."

I mention these Nordic silvaculture experiments to show that a deep cultural appreciation of the boreal forest does not invariably lead to sustainable forestry practices. Sweden and Finland continue to carry out these misguided silvaculture practices in an attempt to supply their forest industries with enough trees to operate mills at full capacity. Some experts predict that it cannot continue at this rate for much longer and that many mills will have to close.

On the other hand, the deep appreciation of the boreal forest evident in these countries provides the public involvement and support needed to restructure these forest industries on a more sustainable basis. Sweden's National Forest Act, for example, enforces a maximum annual harvest quota for forests. Its pulp and paper companies are leaders in the recycling of used paper as part of their forest products. And it is developing effluent-free paper mills. Because of these and other reasons, I remain optimistic about the future management of the boreal forest in Sweden and predict that over the next several decades Swedish forest industries will undergo widespread collapse and then massive transformations. Following this painful period, Sweden will emerge as a world leader in a new type of forestry, one that successfully integrates many different uses of the taiga and one that is truly sustainable, not only in an economic sense but also from an ecological perspective.

The Russian Perception

The traditional Russian perception of the taiga probably comes the closest to the typical Canadian attitude of viewing the boreal forest as remote lands of questionable value. However, the Russian perception of the boreal forest has changed dramatically during the past century. I can only give an overview of this topic here, but I suggest several books that treat it in more detail. To understand how Russians have changed their perception of the boreal forest, we must start in the sixteenth century and learn about the czar's policy for Siberia as a "giveaway kingdom."

In 1547 Ivan the Terrible assumed the imperial title of czar, reigning over a forced assemblage of small Russian states, with Moscow as its proclaimed capital. In 1574 Czar Ivan invited the leaders of the powerful merchant clan, the Strogonovs, to Moscow to promise them a deed to all of western Siberia provided they would conquer the area at their own expense and establish czarist rule throughout the land. The Strogonovs accepted, carrying out the challenge with zealousness and mercantile efficiency. In 1581 the reigning Strogonov patriarch organized an expedition of Cossack mercenaries under the leadership of Yermak, known as The Conqueror of Siberia. By the summer of 1583 all Siberian lands had become part of Russia. The czarist approach for controlling Siberia through economic giveaways was then practiced for more than 300 years. In many ways the approach must be viewed as a great success. The loyalty of the czar's frontiersmen was unfaltering—the Siberian Cossacks were among the last forces to submit to Soviet rule during the Great Revolution.

During the Soviet era and continuing to the present, a very different view concerning the Siberian taiga has emerged among Russians. During the twentieth century Siberia became the foundation of the Russian economy, providing—through slow and painstaking efforts—many of its required resources. On more than one occasion, Siberia rescued the Soviet economy from collapse, an economic bankruptcy that the West quietly but deliberately tried to orchestrate. These events permanently transformed the perception of Siberia among Russians. Let's look at these events more closely.

During the 1920s and 1930s Western countries organized against the new Bolshevik government in an effort to break the communist regime. One of the programs they put in place was a boycott that prevented any corporation from selling the Soviet Union certain scarce resources that it needed for development. This boycott was an attempt to control the crucial natural resources that Soviets needed to buy from other countries in order to develop its petroleum reserves. Without oil and gas, the Soviets could not develop or operate their heavy industries or transportation systems. One of the net results of this boycott was a desperate search throughout the vast regions of Soviet Russia for the needed resources, and it was often in Siberia where these crucial resources were found. As a result of Siberia providing these resources time and time again, the area emerged as a vital and productive part of the Soviet economy.

These international boycotts are illustrated by the West's monopoly of industrial diamonds. As late as the mid 1950s, diamond deposits were unknown anywhere within the Soviet Union. By tightly regulating the supply of industrial diamonds that reached the Soviet Union, Western governments believed they could control the rate of industrial development throughout the USSR. Consequently, Western governments and diamond conglomerates cooperated closely to ensure that only a small supply of industrial diamonds ever reached the Soviet Union. For four decades the strategy worked. And for four decades the Soviets searched intensely and in vain for a source of diamonds within their borders.

Then in the summer of 1954, the breakthrough came. Larissa Popugayeva, a young geologist from Leningrad, made the first major discovery of diamonds, and she discovered them in the remote Siberian region of Yakutia. Since that landmark discovery, a number of ancient diamond pipes (extinct volcanic cones) have been located in the Lena Basin in central Siberia. Development was rapid. The city of Mirny was founded to develop these diamond deposits. Its construction began in December 1955, and today it is a city of 40,000 people and the diamond capital of Russia.

Siberia has thus been transformed from a giveaway land or a land of prisoners and suffering into Mother Siberia. This is not to say that there has not been incredible mismanagement of the natural resources of the Siberian taiga. Massive mistakes were made under the Soviet regime, and they have only worsened under the current

Russian regime. If Sweden can be accused of burning the furniture to keep warm, Russia is dangerously close to running out of furniture. Consider the Maritime Territory of Khrebet Sikhote Alin', in eastern Siberia. Approximately the size of Korea, it is opposite northern Japan. In an attempt to stimulate economic development in this cash-strapped region, Russia permitted Japanese-owned forest companies to burn and clear-cut massive areas, with very few requirements concerning reforestation. Every year approximately 740,000 acres (300,000 ha) of forest are lost to fires and 692,000 acres (280,000 ha) are clear-cut. It is a region headed for environmental disaster.

Conditions have definitely worsened since the breakup of the USSR. In his book *Vanishing Halo,* Daniel Gawthrop summarizes a number of conservation issues and evaluates management practices in circumpolar locations such as Alaska, Canada, and Scandinavia. Concerning Russia, he states:

> Much of Russia's out-of-control development during the 1990s can be attributed to the pressures of an ailing economy following the collapse of the Soviet Union. The Russian government, strapped for cash, was eager to capitalize on the country's natural resources. Selling them off to foreign interests was the quickest and easiest way to generate revenue. . . .
>
> Under pressure by the World Bank and the International Monetary Fund, politicians have found that the temptation to sell off the country's resources is too great to resist. "Western collaboration and investment, of course, are needed," a former economic adviser to the Russian government told *Time* magazine. "But the Russians have been preoccupied with the fight for the spoils. No one is thinking of long-term strategy."

As important as these topics are, reviewing the mismanagement of the taiga or describing the sizable challenges that Russia faces in the future are subjects largely beyond the ken of this book. The main point that I wish to make is that the perception of Siberia by most Russians has been transformed during the past century from a region of chains and prisoners into Mother Siberia. Siberia supplies the Russian economy, and as a result Siberia is still viewed as a land of opportunity, even during these terribly difficult times in Russia.

Taking Stock

We have surveyed how the taiga is perceived by several cultures—Canadian, aboriginal, Swedish, and Russian. Even though the taiga is ecologically similar in its circumpolar sweep, it is perceived in a different manner by each of these cultural groups. The Koyukon people view the boreal forest as a world of spiritual interrelationships, and they harvest its plants, animals, and physical resources with a great deal of respect. Swedes view the boreal forest as a national treasure, the fountainhead of their culture. Russians view the taiga as the foundation of the Russian economy. Given that most northern countries and First Nations perceive the taiga in these positive ways, it is puzzling that so many Canadians continue to view the boreal forest as a vast, unproductive land.

It seems clear to me that the management of Canada's taiga is not going to improve significantly until we perceive the taiga in a better light. Not much progress is going to be made as long as a majority of Canadians view the boreal forest as an economic wasteland instead of a land to be valued, cherished, and carefully managed. When Canadians understand that their boreal forest is an important part of a unique circumpolar biome, the northernmost forest on Earth, then it will be possible to mount an effort to change the way we treat and manage this threatened snow forest.

There are signs that this monolithic misunderstanding of the boreal forest is beginning to crumble. Shallow and uncritical attitudes of the taiga are being challenged. Coalitions between aboriginal groups and environmentalists, as well as the work of writers such as David Suzuki, Daniel Gawthrop, Jamie Bastedo, and others, are waking us up, urging us to think more deeply about the value of the boreal forest. I hope it is not too late.

Canadians have not yet decided what land-use practices are appropriate and how to manage the wildlife populations and other natural resources of the taiga. The lives of many creatures—human and nonhuman alike—depend on the way we proceed. Our decisions will seal the fate of some of the most important lakes on Earth; the largest caribou herds in existence; most of the wolves, grizzly bears, and wolverines found on this continent; the breeding grounds for all whooping cranes left in the wild; as well as the millions of birds that migrate into the boreal forest each summer. They will also influence the future existence of many First Nations cultures. It is a critical time to reflect on the value of this land and how we might transform its economy from one based on overexploitation of resources to an economy that is truly sustainable.

General References

Bastedo, J. 1998. *Reaching North: A Celebration of the Subarctic*. Red Deer, Alberta: Red Deer College Press.

Canadian Senate Sub-Committee on the Boreal Forest. 1999. *Competing Realities: The Boreal Forest at Risk*. Report of the Sub-Committee on the Boreal Forest of the Standing Committee on Agriculture and Forestry. Ottawa: House of Commons.

Gamlin, L. 1988. Sweden's Factory Forests. *New Scientist*, 28 January 1988, pp. 37–54.

Gawthrop, D. 1999. *Vanishing Halo: Saving the Boreal Forest*. Vancouver: Greystone Books, Douglas and McIntyre.

Goldman, M. I. 1987. *Gorbachev's Challenge: Economic Reform in the Age of High Technology*. New York: Norton.

Marsh, J. H., ed. 1994. *The Canadian Encyclopedia*. Edmonton: Hurtig.

Nelson, R. K. 1983. *Make Prayers to the Raven: A Koyukon View of the Northern Forest*. Chicago: University of Chicago Press. (Also see the video series of the same title, produced by the University of Alaska, Fairbanks.)

Pruitt, W. O., Jr. 1991. Importance of Forests and Bogs in the Ecology of Northern Regions. In *Arctic Complexity: Essays on Arctic Interdependencies*, edited by L. Heinemen and J. Käkonen. Occasional Paper 44. Tampere, Finland: Peace Research Institute.

St. George, G. 1969. *Siberia: The New Frontier*. New York: D. McKay.

Valliere, P. 1994. *Change and Tradition in Russian Civilization*. Westland, Mich.: Hayden-McNeil.

Whiting, A. S. 1981. *Siberian Development and East Asia: Threat or Promise?* Stanford, Calif.: Stanford University Press.

13
The Nordic Challenge

Swedes, Finns, Russians, and Canadians perceive their taiga differently, and yet each of these societies has ended up overharvesting the taiga and mismanaging its resources. It makes one wonder if overexploitation of the taiga's resources is the only economic option. Is massive resource harvesting the only feasible approach for supporting an economy in the boreal forest, given today's market-driven global economy?

I believe that there are real, viable economic alternatives to the cut, scrape, rape, and run methods of managing the taiga's resources, and we explore some of these alternatives in this chapter. Not surprisingly, these ecologically sustainable alternatives have been developed by societies that have a positive perception of the taiga. In this chapter we look at the informal economy that exists in many northern First Nations communities. We also look at how certain communities in northern Sweden and Finland are trying to restructure their taiga-based economy along truly ecologically sustainable lines. Finally, we consider how relevant these economic alternatives are for the boreal forest region of Canada.

The Mixed Economy of the North

Challenging our perception of the Canadian taiga is important, but it does not put food on the table. It does not provide people living in the boreal forest with a livelihood to support themselves and their families. This is a basic need, and it can be met in ecologically sustainable ways.

One of the best reasons for restructuring the economy of the Canadian taiga is to protect its true economy. In northern areas there are really two economies coexisting side by side—the formal, wage-earning economy and the informal, or subsistence, economy of living off the land. Thomas Berger, who chaired the important Mackenzie Valley Pipeline Inquiry during the 1970s and has since chaired several public inquiries in both northern Canada and Alaska, has an in-depth understanding of the subsistence economy of the North. In his book *Village Journey,* he explains the importance of this informal economy. He states:

> The traditional economy is based on subsistence activities that require special skills and a complex understanding of the local environment that enables the people to live directly from the land. It also involves cultural values and attitudes: Mutual respect, sharing, resourcefulness, and an understanding that is both conscious and mystical of the intricate interrelationships that link humans, animals, and the environment. To this array of activities and deeply embedded values, we attach the word subsistence, recognizing that no one word can adequately encompass all these related concepts. The subsistence economy is a fascinating economic entity encountered repeatedly throughout northern regions. Many see it as a culture of poverty, but participation in a flourishing subsistence economy is, in fact, one of the drawing cards enticing people to live in remote taiga areas. A social system based on co-operation, it is deeply rooted in cultural and religious traditions.
>
> Indigenous people, perhaps more than any other group, have a deep appreciation for the personal fulfillment and spiritual affirmation such a lifestyle provides. Northern native people often say that the subsistence economy is their culture itself. This intricate mixture of cultural activities, co-operation, and economic enterprise provides a profound sense of belonging to the land. It offers the ultimate connection with nature, a deep sense of membership in the surrounding ecological community—as a participating citizen. Cultural fulfillment comes in ways that no other lifestyle can produce. Life within a subsistence economy allows a practice of values, a deep sense of community, a spirit of receiving and sharing, a quality of health and vigor, and a constantly varying mixture of activity and challenges. Through their subsistence economy, native people participate in a culture that has kept them rooted to their land for centuries.

Thomas Berger understands the relevance of the subsistence economy of the North for its people. In *Northern Frontier, Northern Homeland,* the report of the Mackenzie Valley Pipeline Inquiry, he describes how the formal wage economy and the subsistence economy coexist and often function in mutually beneficial ways. Because both systems function side by side, Berger speaks about the "mixed economy of the North." The symbiosis between these two systems is probably the single greatest economic resource possessed by many northerners.

The Nexus of the Informal Economy

If we wish to use the subsistence economy to improve conditions in the North, we should have a grasp of its components. The subsistence economy of the North is a free-flowing and open-ended set of activities that provides food, shelter, clothing, and other necessities for many families. It adapts to local conditions, so the form it takes varies from area to area and from year to year. A few salient facts illustrate the importance of this informal economy.

A few years ago, Yukon 2000, a government-sponsored task force in the Yukon Territory, focused on the territory's future economic development. As part of its work, it documented the role of "country foods" in the economic well-being of the inhabitants. Country foods are those harvested from nature—meats procured through hunting and fishing, berries picked in the wild, wild eggs and mushrooms collected, and native vegetables preserved. They include candies made from birch syrup, cloudberry preserves, and locally brewed alcoholic beverages. The findings of the task force were impressive: Country foods contribute more than $10 million per year to the Yukon economy. The importance of the subsistence economy to the nutritional well-being of Yukoners is much greater than many people realize.

Because the subsistence economy is nebulous and does not fit into tidy economic analyses, its importance is often underestimated. Economic planners, government officials, and even many northerners undervalue the economic worth of the North's informal economy. In addition, the fashions promoted by the movies as well as media celebrities and sports stars seen on television encourage many northerners, particularly young people, to view their subsistence economy with disdain. On the other hand, many people deeply appreciate the well-being that the informal economy provides. As previously mentioned, living off the land is a major reason that many people give for living in the North.

A subsistence economy is encountered almost everywhere across the circumpolar North. For instance, Torbjörn Lahti from the northern Swedish district of Övertorneå recently explained in an interview with me the value of the subsistence economy for his region:

> In the rural areas of northern Sweden, we have an informal economy as well as a formal one. Here we have hunting and fishing. You don't have to buy potatoes, vegetables, or meat. Most likely you are building your own home in these areas, often with locally produced materials. These are all part of our informal economy. In this area of Övertorneå, you have one of the highest rates of unemployment and lowest per capita average income in any area of Sweden. You should see poverty and widespread unemployment, but you don't see it. Because you are not officially employed does not mean that you are not busy. Here you see people driving relatively new Volvos—people seem affluent. You have the official state picture of the economic conditions of the area based on the formal economy, and then you have reality.

The informal economy in this case consists not only of wild meats and garden produce but also of homes, heating systems, and heating fuels produced by the local society. It also includes the crafts and technical skills that are freely shared in most rural communities. I have a personal indebtedness to the informal economy of the North. When my wife and I built our home by a small lake in northern Saskatchewan, we depended almost completely on the informal economy for technical expertise.

Our piece of land was one of thirty lots opened up just outside Prince Albert National Park. A variety of people—forest workers, national park wardens, and prairie farmers—built their homes and vacation homes by that corner of the lake. Typical of an informal economy, our home and those of our neighbors went up with a great deal of cooperation and sharing of skills, tools, and advice. One neighbor knew how to fell trees, while another knew how to lay concrete. Someone was skilled at wiring, another knew cabinetmaking. Yet another neighbor was knowledgeable about insulation. There was endless swapping of materials and advice, and numerous work arrangements were struck. As our homes were completed, there was a sense of camaraderie, of freedom and self-sufficiency, despite the bone-numbing efforts that accompanied our toils. It was an education that I will always cherish. Not only did it give my family the soundest and warmest home that we shall ever inhabit, but it also gave us the conviction that we will never be homeless. We can always do it again.

Newfoundland's Rural Technology

Newfoundland is perhaps one of the world capitals of subsistence economy. One of the truisms of the mixed economy of the North is that the more hard pressed the formal wage economy of a region, the more resilient and innovative the subsistence economy becomes. The picture most people have of Newfoundland is that of an economically depressed region, and so it is. Unemployment runs at twice the national Canadian average, and per capita income is the lowest in Canada. People lucky enough to be employed often hold only seasonal or part-time positions. Life in such outports is often presented as marginal, consisting of poverty and economic stress.

Actual life in rural Newfoundland, however, is far from grim. For example, Newfoundland has the highest rate of home ownership in Canada. Country foods are highly developed and have given rise to a unique regional cuisine: flipper pie, seal stew, moose potpie, and turr. When rural Newfoundlanders build a home or a fishing dory, they often begin with logs brought straight from the forest. In their backyards there is likely to be a push mill, which consists of two carriages, similar to two horizontal ladders, that slide past each other, combined with a rotary saw blade that might be up to a thirty-six inches (92 cm) in diameter. In Newfoundland these saws are most often driven by an old ship's engine or by an old vehicle mounted on blocks. This portable, one-man sawmill can produce lumber of any dimension desired, even the odd-shaped timbers needed in shipbuilding. I once watched a man patiently cut out of a seasoned balsam fir trunk a two-by-four that was straight and true and twenty feet (6 m) long. That's quite a piece of lumber!

A strong subsistence economy in the taiga is closely tied to healthy ecological processes in its lakes and forests, and this economy must operate within the limits of these ecological processes. Bill Fuller, a retired ecologist from the University of Alberta who did research in Canada's taiga for decades, wisely points out that in northern areas there are severe limitations on the primary production of ecosystems. As a result, there is a limit to the harvest of fish, game, and fur that is sustainable. He suggests that the ability of the renewable food resources (fish, birds, and mammals) to support an expanding human population in the North has frequently been overestimated. Based on calculations for Nunavut and the Northwest Territories, he estimates that the protein-yielding fish and big game populations might support a maximum of double the 1976 human population for that region, a population that may be reached before 2010.

These ecosystem processes, though limited by a harsh climate and short growing season, are the real production mechanisms in a northern economy. For the isolated human communities of the North, they are the producers of their meat and fish and the suppliers of much of their building materials and home-heating fuels. Once the true economic value of the informal economy is realized, resource-extraction proposals, such as mining, timber harvesting, and commercial fishing, will be viewed in a different light. Obviously, renewable and nonrenewable resource extraction has an important role to play and needs to be considered, but it is lunacy to permit those developments to threaten the very basis upon which the subsistence economy of the region depends. The economy of the North will always be a mixed one. Obviously, northerners need at least part-time wage employment to have the cash to buy products that can never be supplied by the subsistence economy. Resource extraction should be designed to meet those needs without threatening the health of northern ecosystems.

Economic planners, resource managers, and elected government officials consistently underestimate the economic worth of the informal economy, and their policy decisions and management actions reflect this misconception. Renewable and nonrenewable resources are often harvested by large corporations in a completely unsustainable manner, and the local residents of the area are left with their informal economy in ruins. Welfare becomes their only economic prospect, and the associated problems of alcoholism, drug abuse, and mental and social breakdown inevitably follow. When the true societal and monetary costs are considered, the sabotage of a sustainable informal economy for a one-shot, cut-and-run, resource-extraction scheme makes no economic sense at all.

It doesn't have to happen like that. The wage component of the North's mixed economy does not have to come from operations that devastate the natural environment, poison its lakes and rivers, and sabotage human communities. Northerners need employment, but the number of wage-paying positions cited as desirable economic targets in planning exercises is often excessive, because planners and government officials lack any appreciation of the economic value of the informal economy. Individuals actively participating in the informal economy should be considered part-time, self-employed entrepreneurs. They provide food, shelter, heating and water systems, and often transportation for themselves and their families using culturally affirming activities. On the other hand, northerners as a group need to believe in and fight for the development of their subsistence economy. It is not a stagnant economic system. It is a system that should be open to innovation. As much effort and research should go into the development of the informal economy as are given to the expansion of the North's formal wage economy.

In fact, very little innovation has been shown regarding economic enterprises in the North, be they formal or informal. Thirty years ago this lack of innovation might have been excusable—roads in the North were primitive or nonexistent; travel costs by chartered airlines were high, and communication systems were poorly developed. However, today large regions of the Canadian taiga are accessible by all-weather roads; there is a network of commercial flights across northern Canada, and satellite and computer communication systems and radio telephones are readily available. Small, computer-based businesses are opening up new economic prospects across the North. It is time to reassess what types of commercial enterprise are now possible in the North and to explore economic development proposals that are compatible with and complementary to the informal economy.

Two Nordic Ecocommunes

In this regard, it is interesting to explore innovations in two northern districts of Sweden and Finland that are experimenting with the development of ecological communes in a taiga environment.

The word *commune* in Swedish or Finnish does not have the same connotation as it does in English. In those countries a commune is equivalent to a county in the United States or to a rural municipality in western Canada. It is simply a division of land with its own local form of government. Sweden is divided into 280 communes. Finland has three kinds of commune, or *kunta*: 43 cities or urban communes, 24 boroughs or semiurban communes, and 482 rural communes. Two of these communes—Övertorneå in Sweden and Suomussalmi in Finland—have declared that they wish to redevelop their local economies along ecological, Earth-sustaining lines. The initiatives were chosen by the rural populations in each commune. Perhaps the best way to understand them is to follow their historical development.

Övertorneå, a region of approximately 914 square miles (2,367 km^2), straddles the Arctic Circle. Traditionally, the people of Övertorneå have made their livelihood by harvesting timber and by agriculture—mainly dairy farming but also vegetable and cereal crops. Their fields and pastures are carved from cleared taiga or they lie on fertile lands along the Torne River, lands that flood each spring and thus that have never been extensively forested.

Suomussalmi is a remote municipality in the central part of Finland; the eastern portion of the Suomussalmi Kunta abuts the international border with Russia. Its eastern portion is a large lake district, while hills dominate the western half. Habitation is concentrated along the water bodies of the region. The human population of the Suomussalmi Kunta peaked in the mid-1960s; but with the increased mechanization of agriculture and forestry operations, it has been declining steadily ever since, largely as a result of people moving to the cities. For example, the number of farms being worked declined from 1,200 in the 1960s to under 500 by the late 1970s. This loss of population has crippled the local economy.

Northern farming in these countries is different from that in North America. The Gulf Stream flows along the entire coast of Norway, warming these Nordic countries and even adjacent parts of the Soviet Union, such as the Kola Peninsula. As a result, these countries lack extensive permafrost, and the land can be farmed much farther north than is possible in North America. In Övertorneå, plants are usually set out in the fields in early June, but if the year is cold, planting may be delayed until after the Midsummer Eve festival (summer solstice). To avoid frost damage, crops are harvested by late August or early September. The growing season is short, but the crops never see darkness. The produce that is harvested is sold in cities and towns throughout northern Sweden, Norway, and Finland.

Övertorneå's ecological orientation is inherent. On its best agricultural lands, those adjacent to the Torne River, crops have been raised for over 3,000 years without the aid of chemical fertilizers. Farmers in this area have always improved the productivity of the soil organically, and it is further renewed every year by the springtime floods, which bring fresh sediment and minerals. As one ascends from the Torne River valley, the soil changes to the typical spodosols (podzols) of the taiga, and farming becomes impossible. At higher elevations, agriculture gives way to reindeer husbandry and forestry.

Birth from a Recession

During the 1930s and 1940s, the district of Övertorneå prospered. Agriculture was at a peak, with new areas being cleared and put into production each year. Cooperative farming unions were very active. During the 1950s and 1960s, conditions changed. Sweden went through massive postwar industrial development, and its national economy expanded rapidly. Many people left the rural areas and migrated southward to urban areas with greater job opportunities and economic prosperity. Large-scale industrial growth continued until 1978, when suddenly Sweden entered a serious three-year recession.

The recession drove people out of the cities and industrial centers and back to rural areas such as Övertorneå. Why did Swedes return to their rural roots when they were faced with an economic crisis? Swedish sociologists suggest that people felt they would be better able to cope with hard economic times if they lived in a rural area or small town, where they could rely on the informal economy. When the recession ended in 1982, many people decided that they preferred the quality of life in rural communities. In Övertorneå it was important to the inhabitants that their families could speak the dialect characteristic of the region and that their children could learn some traditions unique to its culture.

Nevertheless, people were pessimistic about the future of these rural regions. They saw that in areas such as Övertorneå opportunities were limited and economic prospects bleak. They believed that this was particularly true for young adults, who, they felt, would have to move to large cities to be educated and to find employment. In Övertorneå, the 1978 population was 6,000, a reduction of nearly 40 percent from the peak of the 1950s. Throughout the early 1980s emigration continued. Residents saw the population aging, and they felt that their economy was dying. Encouraged by strong national support in the form of a task force called The Municipalities and the Future, the commune government began to consider new options. They were led in this by Torbjörn Lahti, a young, energetic social scientist who became leader of Övertorneå's The Future Project and later the director of its ecology foundation, Stiftelsen Ekotopen, both of which were instrumental in guiding Övertorneå on its fascinating restructuring.

About the same time in Finland, Suomussalmi, concerned about the rate at which local farms were being abandoned, began to contemplate declaring itself an ecological commune. A major conference was held in 1983 in Ammarsaarii, the largest town in this Finnish commune, resulting in the launching of the organization Suomussalmi Ekokunta, whose objective was to offer the region's farmers new alternatives and to revitalize the spirit and the economic prospects of farming families. As it happened, Nils Unga, a Swedish economist who had been born and raised on a farm in the Torne Valley and who had strong feelings about the situation in Övertorneå, organized a group from that commune to attend the Suomussalmi conference. The enthusiasm of that gathering gave birth to the Övertorneå Ecocommune.

During the next year, Torbjörn Lahti and his coworkers in The Future Project held a series of meetings in the villages and towns of Övertorneå, inviting local farmers, forest workers, villagers, and merchants to explore the possibilities of an ecological commune. Lahti said:

> We forced nothing on the local people. We simply presented the challenge that if Övertorneå were to become an ecological commune as a way of rejuvenating its economy, what would this mean to agriculture? What would this mean to forestry? What would it mean to the cultural activities of the area? Slowly each alternative was explored in turn, always with an eye to what new economic initiatives could be started from this new dedication.

The Future Project helped many communities in the commune—from the small city of Matarengi to the isolated forest village of Aapua—to form study groups of individuals who were enthusiastic about the commitment and who wanted to brainstorm the implications of an ecological commune. These study groups were among the main movers of the ecological commune and still develop new ideas at the grassroots level. Let's examine the particular weave of rural undertakings that have been brought together in Övertorneå, beginning at the forest hamlet of Aapua.

Building an Ecological Commune

Aapua, located on a high, forested plateau in the northwestern corner of the Övertorneå Commune, is linked to the outside world by a single small road twisting and winding through the taiga. When I visited there, I was surprised by how much Aapua reminded me of small communities scattered across the Canadian taiga. So many things about this remote Nordic community made me feel at home—the many owner-built houses, each in a different stage of construction, each with its individual quotient of graying lumber revealing much about the family living there; the quiet, shy manner of the people; the spruce-pine forests. On the other hand, the scattered herds of semidomesticated reindeer grazing in the surrounding forests reminded me that I was not in northern Canada. Agneta Enebro showed me around Aapua and helped me with the language, so that soon I was learning the important role that Aapua had played in the development of the ecological commune.

In a very real sense, the inspiration for the idea came from Aapua. The forests around this village have always been rich with berries, a resource heavily utilized by the residents. In fact, many inhabitants of Aapua earn two months of their annual income from harvesting and selling the various types of berry that grow in their forests. During the late 1970s the Swedish National Board of Forestry and forest companies began a program of aerial herbicide spraying to thin the forests, and Aapua was chosen as one of its experimental sites.

There was strong local reaction against the spraying operation, not only because of its effects on the berries but also because it threatened the few forest-management jobs that still existed in the community. For six weeks that summer, the residents of Aapua occupied the forest day and night, preventing the large forest companies from completing their planned aerial experiments. The forestry companies gave up their attempts, and the media coverage of this grassroots resistance was so damaging to the companies' public image, as well as to that of the National Board of Forestry, that they never again tried chemical spraying. Many people credit this environmental standoff with igniting the spirit of the ecological commune.

Today, this quiet village and the nearby village of Svanstein are experimenting with an alternative form of forestry, locally known as the sheep-instead-of-herbicides project. The forests surrounding these communities are being thinned by grazing sheep. The animals selectively browse the broadleaf seedlings, saplings, and shrubs, the very ones the herbicides were meant to kill, while leaving the conifers relatively untouched. The sheep are there for another important reason—they provide wool, the main material used by the Aapua Weaving Cooperative. This cooperative was started by the women of Aapua as a means of producing natural products and providing meaningful employment for themselves. They sell what they produce—yarns dyed using native plants and a line of woven garments, tablecloths, and linens—at craft shows throughout Sweden and through an international mail-order service. The weaving cooperative founded a day-care center and a preschool, and further projects are planned. In its own quiet way, Aapua has contributed much to the idea of the ecological commune of Övertorneå.

In the village of Rantajärvi, twelve miles (20 km) southeast of Aapua, a group of farming families banded together to develop a heritage camp as their contribution to the ecological commune. The area needed new economic opportunities, and the families wanted to draw attention to their northern culture and its farming way of life. They envisioned a summer camp that would offer children of families who had moved away from the area an opportunity to return and learn about the heritage and culture of their parents and grandparents. The youngsters would live and work on a farm for a couple of weeks in July, learning traditional skills and handicrafts and something of the regional dialect of Övertorneå, a Finnish language dialect with strong Swedish influences. This summer camp has run successfully since 1984 and now receives applications from more young people than it can accommodate.

An important—perhaps the most important—effect of the heritage camp has been on the people of Rantajärvi. As a result of the camp, their attitudes have once again become optimistic. People have begun to move back to the village. Enough families with young children have moved back to save the school from closing. As outgrowths of the camp, other new enterprises have started in the village, for example, a traditional baking hut. In this small log building, the women of the village gather one morning each week to do their baking. Not only has this enterprise made the village self-sufficient in bread, but also, as Kerstin Persson said to me, "It is not just the bread, but the ideas that are born in the baking hut that are important."

Edgar Hietalas was another strong supporter of the idea of the ecological commune. He, like his father before him, owns a large greenhouse operation in the Torne Valley. To him, being part of an ecological commune meant that his greenhouse should produce organically grown, chemical-free produce. Furthermore, he relocated his greenhouse to land that had never been exposed to such chemicals. Later, local farmers formed the Organic Farming Association of the Övertorneå Commune. Their organic produce is bought by the Övertorneå Health Center and is

Inside the baking hut at Rantajärvi in northern Sweden

specially labeled and marketed in nearby cities. The Övertorneå Health Center is one of the most ambitious initiatives of the Övertorneå Commune. It is the single largest purchaser of the organically grown produce of the region, an interrelationship that was no accident. In all of the ecocommune's programs, mutual support has been carefully thought out and developed.

The Health Center was the inspiration of Allan Lehto, one of the region's most successful entrepreneurs. He convinced the commune government to invite Maija Ruisniemi, a nurse from Finland, to come to Övertorneå to direct the development of a holistic health center. Maija is much more than a highly experienced nurse; for years she has been a practitioner of Finnish herbal folk medicine, learned from her parents and grandparents. She is a shy, quiet woman who is extremely dedicated to her patients and to her work. She says, "I simply allow the results to speak for themselves." And those results motivate her patients to sing the praises of the Övertorneå Health Center all over Scandinavia.

The center, almost exactly on the Arctic Circle near the village of Svanstein, lies in a beautiful forested valley. A renovated residential school has been converted into a rustic and attractive health village. An aquatic complex, with pools of different temperatures and salinity, is an important part of Maija's treatments for certain ailments. I arrived at the center in time for the midday smörgasbörd, an organically grown vegetarian banquet served in the beautiful pine-paneled dining room of the main lodge. The patients of the center, who suffer from a variety of health disorders—allergies, arthritis, obesity, vascular problems, ulcers, and gastrointestinal ailments—were enjoying their meals according to the exact instructions conveyed to them during their regular consultative sessions with Maija. As we ate, Maija explained to me: "It is not just the diet but the totality of the program that the Health Center offers that helps to improve people's conditions. The organic food is a large part of it, as is the total freedom from all forms of caffeine and refined sugars and toxic substances often found in common foods and meats. But the rest and physical exercise that are also prescribed have a great deal to do with it."

Each person who comes to the center goes through a rigorous assessment procedure with Maija, following which she tailors a program for them. It might, for example, require a fast lasting up to seven days, during which the patient consumes only fruit juices and specific herbal teas in an effort to rid his or her system of detrimental toxic substances. A physiotherapist and a physician are on call twenty-four hours a day. The Övertorneå Health Center is not alone in its holistic approach. An increasing number of Swedish physicians support the view that certain ailments should be treated not only medically but also through diet and exercise. The Övertorneå Health Center is the northernmost holistic health center in Sweden, and from all appearances it is highly successful.

These are some of the enterprises by which Övertorneå's ecological commune is rejuvenating the economy of the region. The people who live in the commune are exploring others. Eventually Övertorneå's specialized agricultural products will be marketed across Scandinavia. They foresee a line of herbal teas and other health products produced from native plants. They also envision an expanded tourist industry developed from the increasing number of people who come to this northern district out of curiosity about the Övertorneå ecocommune.

Suomussalmi's Food Park

While all this was taking place in Sweden, in Suomussalmi, Finland, the Ekokunta was striving to rejuvenate that region's agricultural economy. Kauko Heikurainen, one of the movement's leaders, explained, "We started the Suomussalmi Ekokunta as a way to explore new initiatives to keep the farms living in our countryside. We wished to stem the number of farms that were dying and lower the decrease of our rural population." To achieve these objectives, the Ekokunta has formed the Food Park, an information and resource center promoting new agricultural enterprises. The Food Park has a number of programs, two of which are a berry factory and a honey industry. Recently the Food Park expanded into the marketing of freshwater fish, harvested from the region's lakes, and of smoked reindeer meat. Let's look at two of these enterprises.

The country surrounding Suomussalmi is rich in wild berries—predominantly blueberries and cloudberries. The income from collecting these wild berries is important to much of the population, especially retired people, unemployed people, and rural residents. This money is not subject to income tax, a deliberate action taken by the Finnish Parliament to increase the incentive for berry collecting. Some families earn up to 20,000 Finnish markka (approximately $5,000 Canadian or $3,000 U.S.) a year through harvesting wild berries. The Suomussalmi berry factory was developed to extend this important component of the informal economy. During a recent season, the berry factory purchased approximately 50,000 tons (51,000 metric tons) of wild berries from the area's residents. Because the berries brought into the factory are mixed with leaves, twigs, and other forest debris, they are immediately frozen so that subsequent mechanical sorting, handling, and cleaning won't injure them. At present, the factory sells only frozen berries in bulk; however, there are plans to develop more labor-intensive products, such as preserves, juice concentrates, and liqueurs.

The clear-cuts around Suomussalmi abound in fireweed, the wildflower so common in clear-cuts and burned-over areas throughout much of the circumpolar taiga. When the people of the region looked about them for underutilized resources, these massive swathes of magenta fireweed caught their notice. Could simple wildflowers be of any economic value? Lauri Ruottinen, a member of the Ekokunta, convinced them to try honey production, using fireweed as the principal nectar source. The new enterprise has not been without its setbacks and technical difficulties. The deep cold of winter necessitates overwintering the beehives indoors, in basements or insulated huts. During early spring the bees need a special energy-rich food source, so when they are first brought out in April, the hives must be placed near flowering willow bushes or provided with pollen mixed with sterilized honey. During May they are moved to their summer locations in clear-cuts, where the bees can utilize in turn the flowers of blueberries and raspberries and finally the abundant and long-flowering fireweed.

Ruottinen calculates that approximately seven acres (3 ha) of good fireweed growth can support five to eight hives and that such a clear-cut will be able to produce honey for ten to fifteen years before the new forest begins to shade out these dense growths of wildflowers. The Suomussalmi Beekeepers Association has found that it takes approximately two years to develop a productive honey operation and that, while it takes 100 hives to produce a livable income, some families in the association are able to maintain over 200 hives in the clear-cut areas that surround them. These two industries—honey and wild berries—are new agriculture enterprises in Suomussalmi and are definitely growth industries. They illustrate what can be accomplished when innovative thinking is used to develop the informal economy into a contemporary lifestyle.

The Future of the Canadian Taiga

I believe the approaches taken in Övertorneå and Suomussalmi have great applicability to communities across the Canadian taiga, although the Canadian version of an ecological commune might necessitate different initiatives. It could emphasize country foods or living off the land or an affirmation of native cultures, but the principles will be the same: a rejection of cut-and-run resource-extraction operations, which leave the land devastated and the ecological productive processes in ruins. In its place would be a combination of sustainable development activities compatible with the ecology of the taiga.

Elements of this are already in place. For example, the wild rice industry has opened new economic opportunities across the taiga regions of the Prairie Provinces. In northern Saskatchewan, the industry has experienced impressive growth over the past twenty years. Wild rice is planted on slow-moving creeks and streams, mostly in the Canadian Shield areas of the North. As it is at present practiced, this northern agriculture uses no fertilizers or herbicides. The slow movement of water through the canelike stalks provides all the nutrients that the crop needs. In 1975 only a handful of people were growing wild rice in northern Saskatchewan; a decade later, a total of 1,800,000 pounds (816,000 kg) was produced. Furthermore, the Wild Rice Growers Association of Northern Saskatchewan has developed a processing plant in La Ronge, which cleans and polishes the rice and provides a number of year-round jobs for residents of the area. That type of sustainable enterprise is in stark contrast to the oversized pulpwood clear-cuts or mining operations that threaten the ecological processes upon which the informal economy of the taiga depends.

Another idea whose time has come is northern aquaculture; that is, trout farming in the thousands of small ponds and kettle lakes so common throughout the taiga. These bodies of water often teem with the larvae of mosquitoes and other aquatic insects. Trout fingerlings stocked into these ponds and lakes often flourish on the abundant insect larvae and can be harvested when they reach a certain size. The product? Pan-sized trout offering a chemical-free, protein-rich, low-saturated-fat meal. In addition to the self-employment opportunities provided for the trout farmers, local jobs are created to process and market the fish.

Other ideas could be developed. In the shops of Stockholm a small jar of cloudberry preserves costs the equivalent of $28 U.S. Cloudberries grow in bogs throughout the northern United States and Canada and are often harvested locally as country foods; however, their economic potential, along with that of other wild northern berries, has received only limited attention.

In the Canadian taiga there could be a thriving industry emphasizing ecological tourism and environmental education. As a society, North Americans have lost their connection to the land. Many environmentalists believe that this is a fundamental cause of our contemporary environmental dilemma. We must stop viewing the natural world as a collection of resources placed here for our use. We are bona fide members of our ecological communities, but until our worldview and lifestyles begin to reflect this reality, we will continue to act as if we have dominion over nature and thus continue to be the unconscious architects of our own demise.

Across the continental span of the Canadian taiga, the indigenous cultures are still very much intact. We can ask them to help us develop relationships with the Earth. The subsistence economy and lifestyle of aboriginal people of the North are expressions of their rootedness to the land. Their culture is their lifestyle, and their lifestyle is intricately connected to the land. There is an opportunity to develop many heritage enterprises from cultural camps that would welcome natives and nonnatives alike to in-depth programs examining the traditional ecological knowledge of this continent's first peoples.

An ecological commune in the Canadian taiga should weave together many different strands. It should be a clustering of diverse initiatives. A sustainable lifestyle in the boreal forest has always necessitated harvesting a range of species, each in its appropriate season. This mixture of activities is a necessary constraint, dictated by the restricted productivity of the taiga. This attribute of northern ecosystems cannot be ignored. It is an important guiding principle that all northern societies have had to affirm. The ecological communes of Övertorneå and Suomussalmi are simply contemporary manifestations of this principle. Northern development must stop focusing on large-scale exploitation of single resources. A gentler approach to the land is needed. The development of the North must be based on the integrated harvesting of diverse resources, each utilized to an appropriate extent, each in its own season, if we want our human communities in the North to be truly sustainable.

Let's hope that in the future the taiga region of Canada will become increasingly a land of ecological and cultural experiences, reconnecting us to the land and to the real community to which we belong and depend. Let's hope that these two sides of the economy develop in concert and develop in ways compatible with the fragile, cold-limited ecology of the northernmost forest on the face of the Earth.

General References

Berger, T. R. 1985. *Village Journey: The Report of the Alaska Native Review Commission*. New York: Hill and Wang.
Berkes, F., C. Folke, and M. Gadjil. 1995. Traditional Knowledge, Biodiversity, Resilience, and Sustainability. In *Biodiversity Conservation*, edited by C. A. Perrings, K.-G. Maler, C. Folke, C. S. Holling, and B. O. Jansson. Dordrecht, Netherlands: Kluwer Academic Publishers.
Brody, H. 1981. *Maps and Dreams*. Vancouver: Douglas and McIntyre.
Övertorneå Kommun: Tungshusvage 2, S-95 785, Övertorneå, Sweden. http://www.overtonea.se.
Scott, F. D. 1988. *Sweden: The Nation's History*. Carbondale: Southern Illinois University Press.
Suomussalmi Kunta: PL 40, 8960, Ammansaari, Suomi, Finland. http:www.suomussalmi.fi.
Winterhalder, B. 1981. Foraging Strategies in the Boreal Forest: An Analysis of Cree Hunting and Gathering. In *Hunter-Gatherer Foraging Strategies: Ethnographic and Archeological Analyses*, edited by B. Winterhalder and E. A. Smith. Chicago: University of Chicago Press.
Wirén, E. 1983. *Every Man's Forest*. Stockholm: Swedish Environmental Protection Board.

14

Places to Visit in Canada's Boreal Forest

It is one thing to read about the boreal forest but quite another to experience it. This chapter is an invitation to experience firsthand the plants, animals, and ecology of Canada's snow forest. Included is a list of parks and protected areas in or near the Canadian taiga (see the map in Chapter 1) as well as museums and cultural centers that interpret this ecological region. The list is not exhaustive; it is meant to assist you in planning a trip to the great northern forest of Canada. Excellent information about all of the National Parks and National Historic Sites of Canada is available on the Parks Canada website, http://parkscanada.pch.gc.ca.

There are also provincial and territorial parks, wilderness areas, historic parks, and regional museums to explore as you travel through the Canadian taiga. For each of the ten provinces and three territories of Canada listed below, you will find a website for general tourist information, contacts for national parks and a few national historic sites that are in or near the taiga, and websites for museums and cultural centers that interpret the boreal forest. Two other web sites that you may find informative are World Heritage Sites in Canada, http://www.cco.caltech.edu/~salmon/wh-canada.html; and CHIN-Guide to Canadian Museums and Galleries, http://www.rcip.gc.ca/museums/e_museums.html. The information listed below is organized east to west, first by province, then by territory.

Newfoundland and Labrador

The most easterly province of Canada comprises the island of Newfoundland in the North Atlantic and adjacent mainland Labrador. Labrador is underlain by the Canadian Shield, while most of Newfoundland lies in the Appalachian Mountains geological region. As with each province and territory, their geology is complex and challenging. Southern Newfoundland supports closed boreal forest; northern Newfoundland and most of Labrador supports open boreal forest. Higher elevations exhibit the forest-tundra ecotone and the tundra biome. For general tourist information: http://www.gov.nf.ca/tourism/.

- Gros Morne National Park (western Newfoundland). P.O. Box 130, Rocky Harbour, Newfoundland, Canada A0K 4N0. Telephone: (709) 458-2417. TDD (device for the deaf): (709) 772-4564. Fax: (709) 458-2059. E-mail: grosmorne_info@pch.gc.ca
- L'anse aux Meadows National Historic Site (northwestern Newfoundland). P.O. Box 70, St-Lunaire-Griquet, Newfoundland, Canada A0K 2X0. Telephone: (709) 623-2608. Fax: (709) 623-2028.
- Terra Nova National Park (eastern Newfoundland). Glovertown, Newfoundland, Canada A0G 2L0. Telephone: (709) 533-2801. Fax: (709) 533-2706. TDD: (709) 772-4564. E-mail: atlantic_parksinfo@pch.gc.ca.

Nova Scotia

This maritime province consists of a large peninsula and many islands jutting out into the North Atlantic. It is part of the Appalachian Mountains geological region. The province mainly supports the northern hardwood and conifer ecotone, but outliers of the boreal forest occur on Cape Breton Island and in southern areas of the province. For general tourist information: http://www.parks.gov.ns.ca/.

- Cape Breton Highlands National Park (northern Nova Scotia). Ingonish Beach, Nova Scotia, Canada B0C 1L0. Telephone: (902) 224-3403 or (902) 224-2306. Fax: (902) 224-2306. TDD: (902) 285-2691. E-mail: atlantic_parksinfo@pch.gc.ca.
- Kejimkujik National Park (southern Nova Scotia). P.O. Box 236, Maitland Bridge, Nova Scotia, Canada B0T 1B0. Telephone: (902) 682-2772. Fax: (902) 682-3367. E-mail: Kejimkujik_Info@pch.gc.ca.

New Brunswick

This province occupies the Appalachian Mountains geological region lying northeast of Maine and west of Nova Scotia. Its vegetation is part of the northern hardwood and conifer ecotone, with outliers of the boreal forest. For general tourist information: http://www.new-brunswick.net/new-brunswick.

- Fundy National Park (southern New Brunswick). P.O. Box 1001, Alma, New Brunswick, Canada E4H 1B4. Telephone: (506) 887-6000. Fax: (506) 887-6011. E-mail: Fundy_info@pch.gc.ca.
- Kouchibouguac National Park (northern New Brunswick). 186, Route 117, Kouchibouguac National Park, New Brunswick, Canada E4X 2P1. Telephone: (506) 876-2443. Fax: (506) 876-4802. TDD: (506) 876-4205. E-mail: kouch_info@pch.gc.ca.

Prince Edward Island

An island in the Gulf of St. Lawrence, P.E.I. is Canada's smallest province. It is a low-relief portion of the Appalachian Mountains geological region and one of the most agriculturally productive parts of Canada. Its natural vegetation is part of the northern hardwood and conifer ecotone. For general tourist information: http://www.gov.pei.ca. The boreal forest does not occur on Prince Edward Island.

Quebec

More than 90 percent of Quebec is Precambrian shield. The St. Lawrence Lowlands form the lands adjacent to Quebec's largest river, and the Appalachian Mountains geological region forms the Gaspè Peninsula. The natural regions of Quebec consist of four zones: The northern hardwood and conifer ecotone occurs along the St. Lawrence River and in southern Quebec. As one travels north, the closed boreal forest changes to the open boreal forest and then to the forest-tundra ecotone. These three vegetation types cover more than three-quarters of the province. In the far north of the province the Arctic tundra dominates. For general tourist information: in French, http://www.mef.gouv.qc/fr/parc_que/; in English, http://www.mef.gouv.qc/en/parc_que/.

- Forillon National Park (eastern Quebec). P.O. Box 1220, 122 Gaspé Blvd., Gaspé, Quebec, Canada G0C 1R0. Telephone: (418) 368-5505. Fax: (418) 368-6837. E-mail: parkscanada-que@pch.gc.ca.
- La Mauricie National Park (southern Quebec). 794-5e Rue, P.O. Box. 758, Shawinigan, Quebec, Canada G9N 6V9. Telephone: (819) 536-2638. Fax: (819) 536-3661. E-mail: parkscanada-que@pch.gc.ca.
- Mingan Archipelago National Park (eastern Quebec). 1 303 rue de la Digue, P.O. Box 1180, Havre-Saint-Pierre, Quebec, Canada G0G 1P0. Telephone: (418) 538-3285 in season (June 8 to September 13). Telephone: (418) 538-3331 off season. Fax: (418) 538-3595. E-mail: archipel_de_mingan@pch.gc.ca.
- Musée de la civilisation (Quebec City, Quebec): http://www.mcq.org/.
- Musée du Québec (Quebec City, Quebec): http://www.mdq.org/.
- National Museum of Civilization (Hull, Quebec): http://www.cmcc.muse.digital.ca/cmc/cmceng/welcmen.

Ontario

All of southern Ontario is part of the St. Lawrence Lowlands geological region, while lands adjacent to Hudson and James Bays are geologically Hudson Bay Lowlands. The remainder of Ontario is underlain by the Canadian Shield. The natural regions of Ontario are similar to those of Quebec, with two important distinctions: The southernmost part of Ontario is classified as eastern deciduous forest, and there is only a narrow band of Arctic tundra adjacent to Hudson Bay along Ontario's northern coast. For general tourist information: http://www.mnr.gov.on.ca/mnr/parks/index/html.

- Bruce Peninsula National Park (southern Ontario). P.O. Box 189, Tobermory, Ontario, Canada N0H 2R0. Telephone: (519) 596-2233. Fax: (519) 596-2298. E-mail: bruce_fathomfive@pch.gc.ca.

- Georgian Bay Islands National Park (southern Ontario). P.O. Box 28, Honey Harbour, Ontario, Canada, P0E 1E0. Telephone: (705) 756-2415. Fax: (705) 756-3886. TDD: (705) 756-2415. E-mail: Mike_Bondar@pch.gc.ca.
- Pukaskwa National Park (central Ontario). Heron Bay, Ontario, Canada P0T 1R0. Telephone: (807) 229-0801. Fax: (807) 229-2097. E-mail: ont_pukaskwa@pch.gc.ca.
- National Museum of Natural History (Ottawa): http://www.nature.ca/.
- Royal Ontario Museum (Toronto): http://www.rom.on.ca.

Manitoba

Three-quarters of Manitoba is underlain by Precambrian shield; the southwestern quarter is part of the Interior Plains geological region. Southwestern Manitoba is part of the grassland biome, but as you travel north it changes fairly quickly into the aspen parkland ecotone. More than two-thirds of Manitoba is closed boreal forest, changing to open boreal forest and then to forest-tundra ecotone farther north. Similar to Ontario, there is a narrow band of Arctic tundra along the Hudson Bay coast. For general tourist information: http://www.gov.mb.ca/natres/parks.

- Riding Mountain National Park (southwestern Manitoba). Wasagaming, Manitoba, Canada R0J 2H0. Toll Free: 1-800-707-8480 (North America only). Telephone: (204) 848-7275. Fax: (204) 848-2596. TTY/TDD: (204) 848-2001. E-mail: RMNP_Info@pch.gc.ca.
- Wapusk National Park (northeastern Manitoba). Wapusk National Park, c/o Churchill Office, P.O. Box 127, Churchill, Manitoba, Canada R0B 0E0. Telephone: (204) 675-8863. Fax: (204) 675-2026. E-mail: wapusk_np@pch.gc.ca.
- Manitoba Museum of Man and Nature (Winnipeg, Manitoba): http://www.manitobamuseum.mb.ca/muse.htm.

Saskatchewan

The Canadian Shield forms most of the northern half of Saskatchewan, with the Athabasca sandstone formation occupying the northwest. The southern half of Saskatchewan is part of the Interior Plains geological region. The southern third of Saskatchewan is grassland; it changes into a narrow aspen parkland ecotone and then a broad area composed of closed boreal forest. On the fertile Interior Plains, a fine mosaic of eight to ten different forest types forms the productive mixed woods section of the boreal forest. In the northeastern quarter of the province, the open boreal forest dominates. For general tourist information: http://www.serm.gov.sk.ca/parks.

- Prince Albert National Park (central Saskatchewan). Box 100, Waskesiu Lake, Saskatchewan, Canada S0J 2Y0. Telephone: (306) 663-4522 (general information and visitor center). Telephone ($5.00 fee): (306) 663-4513 (camping reservations). Fax: (306) 663-5424. E-mail: PANP_INFO@pch.gc.ca.
- Royal Saskatchewan Museum (Regina): http://www.gov.sk.ca/rsm.

- Wanuskewin Heritage Park (Saskatoon): http://www.saskriverbasin.ca/wanuskewin/.

Alberta

The province has a small amount of Precambrian shield in its northeastern corner. Alberta is predominantly Interior Plains. The Canadian Rockies, in the southwest portion of the province, are part of the Canadian Cordillera geological region. The southeastern quarter of Alberta is grasslands, changing into a relatively broad aspen parkland ecotone. The remaining three-quarters of the province is northern coniferous forest, characterized by taiga growing on the Interior Plains and montane coniferous forest growing in the mountains (see Chapter 1). For general tourist information: http://www.gov.ab.ca/env/parks/prov_parks.

- Banff National Park (southwestern Alberta). P.O. Box 900, Banff, Alberta, Canada T0L 0C0. Telephone: (403) 762-1550. Fax: (403) 762-1551. E-mail: banff_vrc@pch.gc.ca.
- Elk Island National Park (central Alberta). Site 4, RR 1, Fort Saskatchewan, Alberta, Canada T8L 2N7. Telephone: (780) 992-2950. Fax: (780) 992-2951 or (780) 992-2983. TDD: (780) 992-2993. E-mail: elk_island@pch.gc.ca.
- Jasper National Park (western Alberta). P.O. Box 10, Jasper, Alberta, Canada T0E 1E0. Telephone: (780) 852-6176. Fax: (780) 852-5601. E-mail: jnp_info@pch.gc.ca.
- Wood Buffalo National Park (northern Alberta). P.O. Box 750, Fort Smith, Northwest Territories, Canada X0E 0P0. Telephone: (867) 872-7900. Fax: (867) 872-3910. TDD: (867) 872-3727. E-mail: wbnp_Info@pch.gc.ca.
- Head-Smashed-In Cultural Center (near Fort Macleod): http://www.head-smashed-in.com.
- Provincial Museum of Alberta (Edmonton): http://www.gov.ab.ca/MCD/mhs/ pma/pma.htm.
- Royal Tyrrell Museum of Palaeontology (Drumheller): http://www.gov.ab.ca/ MCD/mhs/rtmp/rtmp.htm.

British Columbia

The northeastern quarter of the province is part of the Interior Plains geological region; the remainder is composed of the Canadian Cordillera. British Columbia is the most mountainous province of Canada. Its vegetation is largely determined by its moist maritime climate and its mountainous terrain. It has a coastal rainforest along the Pacific coast, but there is an arid grasslands region in the rain shadow of the coastal mountains. Much of the remainder of the province is montane coniferous forest, a majority of it being closed boreal forest changing to open boreal forest in the northwest (see Chapter 1). For general tourist information: http://www.swparks.com/canada/british_columbia.html.

- Chilkoot Trail National Historic Site (northwestern British Columbia). Department of Canadian Heritage, Yukon District Office, 205-300 Main

- Street, Whitehorse, Yukon, Canada Y1A 2B5. Telephone: (867) 667-3910. Fax: (867) 393-6701. E-mail: whitehorse_info@pch.gc.ca.
- Glacier National Park (southeastern British Columbia). P.O. Box 350, Revelstoke, British Columbia, Canada V0E 2S0. Telephone: (250) 837-7500. Fax: (250) 837-7536. TDD: (250) 837-7500. E-mail: revglacier_reception @pch .gc.ca.
- Kootenay National Park (southeastern British Columbia). P.O. Box 220, Radium Hot Springs, British Columbia, Canada V0A 1M0. Telephone: (250) 347-9615. TDD: (250) 347-9980. Fax: (250) 347-9615. E-mail: kootenay_reception@pch .gc.ca.
- Mount Revelstoke National Park (southeastern British Columbia). P.O. Box 350, Revelstoke, British Columbia, Canada V0E 2S0. Telephone: 250 837-7500. Fax: 250 837-7536. TDD: 250 837-7500. E-mail: revglacier_reception@pch.gc.ca.
- Yoho National Park (southeastern British Columbia). P.O. Box 99, Field, British Columbia, Canada V0A 1G0. Telephone: (250) 343-6783. Fax: (250) 343-6330. TDD: (250) 343-6783. E-mail: Yoho_Info@pch.gc.ca.
- Royal British Columbia Museum: http://rbcm.rbcm.gov.bc.ca.

Nunavut

This territory in northeastern Canada spans three geological regions: the Canadian Shield forms its mainland portions and Baffin Island; the Arctic Lowlands form many of the southern Arctic Islands, and the Innuitian Mountains form the northern Arctic Islands. Open boreal forest and the forest-tundra ecotone occupy the southern portions of Nunavut, while the Arctic tundra occupies the remainder of the territory. For general tourist information: http://www.nunavutparks.com/. The boreal forest occurs in the southern portion of Nunavut Territory. No national parks have been established there. Nunavut Territory contains several impressive national parks in the Arctic cordillera, for example, Auyuittuq, Sirmilik, and Quttinirpaaq National Parks.

Northwest Territories

The N.W.T. is geologically similar to Nunavut with the exceptions that the western half of the territory is composed of Interior Plains and that the Canadian Cordillera forms its western mountains. The southwestern half of the N.W.T. is closed boreal forest, which changes to open boreal forest and then to forest-tundra ecotone. The northern half of the territory is Arctic tundra. Together with Nunavut, it forms one of the largest expanses of Arctic tundra found anywhere in the world. For general tourist information: http://www. nwttravel.nt.ca.

- Nahanni National Park Reserve (southwestern Northwest Territories). Postal Bag 300, Fort Simpson, Northwest Territories, Canada X0E 0N0. Telephone: (867) 695-2713. Fax: (867) 695-2446. E-mail: Nahanni_Info@pch.gc.ca.

- Prince of Wales Northern Heritage Center (Yellowknife): http://www.pwnhe .learnnet.nt.ca/.

Yukon Territory

Geologically, all of the Yukon is Canadian Cordillera, with a small band of continental shelf next to the Beaufort Sea. The Yukon is the northern extension of the mountainous cordillera that forms all of British Columbia. The Yukon is a mosaic of closed and open boreal forest intermixed with forest-tundra ecotone and Alpine and Arctic tundra—all responding to the Yukon's northern climate and rugged mountainous terrain. For general tourist information: http://www.touryukon.com/.

- Ivvavik National Park (northern Yukon Territory). Box 1840, Inuvik, Northwest Territories, Canada X0E 0T0. Telephone: (867) 777-3248. Fax: (867) 777-4491. E-mail: Gerry_Kisoun@pch.gc.ca.
- Kluane National Park and Reserve (southwestern Yukon). Box 5495, Haines Junction, Yukon, Canada Y0B 1L0. Telephone: (867) 634-2329, ext. 250. Fax : (867) 634-7208. E-mail: Whitehorse_info@pch.gc.ca.
- Vuntut National Park (northern Yukon Territory). Department of Canadian Heritage, Yukon District Office, 205-300 Main Street, Whitehorse, Yukon, Canada Y1A 2B5. Telephone: (867) 667-3910. Fax: (867) 393-6701. E-mail: whitehorse_info@pch.gc.ca.
- Kwaday Dan Kenji (Long Ago . . . People's Place), an interesting cultural camp located on the Alaska Highway near Champagne, Yukon. Contact Kwaday Dan Kenji, Indianway Ventures, Site 3, Comp 1J1, Whitehorse, Yukon Territory, Canada. Phone: (867) 667-6375.
- Yukon Beringia Interpretive Center (Whitehorse): http://www.beringia.com.

Good and Safe Travels

I hope that the information provided here assists you in planning your trip to Canada's boreal forest. Enjoy your travels and explorations of the taiga. I hope that this book increases your enjoyment and appreciation of the snow forest, the northernmost forest on Earth.

General References

Lynch, W. 2001. *The Great Northern Kingdom: Life in the Boreal Forest*. Ontario: Fitzhenry and Whiteside.

Maybank, B., and P. Mertz. 2001. *The National Parks and Other Wild Places of Canada*. London: New Holland.

McNamee, K. A. 1994. *The National Parks of Canada*. Toronto: Key Porter Books.

Index

acidification, 11, 66, 94, 101, 104, 106–7, 110, 124
acid-loving plants, 64, 104, 107
acid rain, 23, 115, 124–27, 135, 142
Adirondack Mountains, 9
air mass, 13–14
Alaska, 2, 7, 9, 10, 21–24, 49, 70, 79, 87, 92, 96, 104, 131, 144
Alberta, xiii, 7, 9, 21, 33, 44, 85, 86, 92, 110, 119–20, 143, 169
algae, 28, 117
Ammarsaasii, Finland, 157
Appalachian Mountains, 32, 33, 87, 165, 166, 167
aquaculture, 110, 163
aquatic ecosystems, 116–18
archaeological site, 22, 113
Arctic Circle, 10, 156, 161
aspen, trembling, 3, 38, 40, 43, 50, 53, 60, 61, 63, 89, 94, 104
aspen parkland, 2, 9, 168, 169
Athabascan, 7, 44
atmosphere, 20, 111, 116; ozone layer of, 20
aurora borealis, ix, 115

bears: American black, 18, 115, 123, 145; European brown, 145; grizzly 15, 19, 20, 150
beaver, 15, 16, 69, 70, 73, 83, 116, 121
bees, 47–48, 162–63; beekeepers, 162–63
beetles, spruce bark, 22
Berger, Thomas, 7, 151–52
berries, 3, 60–61, 115, 158, 162–63
biomes, ix, xi, 1, 150
birch: dwarf, 50, 64; paper or white, 2, 43, 50, 53, 60, 61, 63, 66, 67, 94–95, 115

birds, ix, 111, 119, 120; long-distance migrants, 19; short-distance migrants, 15, 19. *See also* grouse; owls; passerines; warblers
bison, 19, 77, 120
bogs, 45, 103–5, 107–8, 146–47; ombrotrophic, 102
Boonstra, Rudy, 92–93, 96, 99
boreal forest, 35, 42–44, 46, 48, 51, 87, 130, 166; defined, 1–3; northern (open), 4–5, 165, 167, 168, 169, 170, 171; southern (closed), 4–5, 11, 60, 63, 64, 165, 167, 168, 169, 170, 171
botulism, 133
boulder field, 45–46
break-up of lakes or rivers, 39–41, 119–20, 122
British Columbia, 7, 9, 28, 110, 120, 143, 169–70
Bryant, John P., 94–96, 98–99

cabin, family, xi, 129, 146, 153–54
Canadian Shield, xi, 5, 29–31, 32, 35, 37–38, 110, 112, 125–26, 141, 163, 165, 167, 168, 170. *See also* Precambrian shields
carbon sink, 111–12, 147
caribou, 15, 16, 19, 77–79, 113, 120, 150
caterpillars, tent, 86–87
Cenozoic Era, 28
Chekhov, Anton, 1, 38
chickadee, 18
Chipewyan, 7, 105, 120
climate change, 20–24, 34, 43, 111, 123. *See also* global warming
climax communities, 59–60, 63, 65–67
cold hardiness, 12, 116
communes, ecological, 156–64

communities, northern, 23, 121, 151, 155–56, 157, 158–61
cones: crop failures of, 18, 98; serotinous, 43, 54–58, 60
conifers, ix, 2, 9, 10–12, 41, 44, 63
continental drift, 29–30, 34
coprophagy, 89
Cordillera, Canadian, 2, 5, 32, 33
cougars, 15, 19
country foods, 153–54, 162–63
cratons, 29
Cree, Lubicon, xiii
Cree, Woods, 7, 39–40, 46, 80, 105, 108, 112–14, 115, 120, 129
cycles; four-year, 18, 91; ten-year 18, 85–93, 96–99. *See also* population cycles

dams, hydroelectric, 23, 120–21, 143
decomposition, 11, 64, 65–67, 94, 104, 108, 111, 146
delta, river, 5, 22, 51, 119–20
detritus, 66
diamonds, 148
dinosaurs, 28
dogs, sled, 80

economy: ecologically sustainable, xiii, 112, 126–27, 151, 155; formal (wage), 151–56, 164; informal (subsistence), 151–56, 164
endangered species, 19–20
energy budget, 10, 70, 83
Environment Canada, 5, 19–20
evapotranspiration, 42, 101
evolution, 40–41, 47–48, 54–57, 66–67, 88, 95, 101, 106, 108, 116
Experimental Lakes, Ontario, 22, 123
experimentation, 67, 71, 89, 95, 96–97

farming, organic, 156–57, 159–60
feather moss, 63–65
fens, 102–4; ribbed, 105
Finland, 12, 79, 87, 144, 146–47, 156–57, 159, 162–63
Finnish herbal folk medicine, 161
fir, balsam, 9, 12, 44, 50, 61, 63
First Nations, 7, 144, 149–50, 151–52, 164. *See also* indigenous people
fishing: commercial, 133, 135; sports, xi, 115, 123–24, 133, 141–42
fish populations, 22–23, 110, 118, 120, 121, 122, 123, 125–26, 154–55
flammable characteristics of trees, 51–53, 56–57, 65–67
forest fires, x, 11, 22, 41–43, 49–51, 63, 66–68, 95, 111; crown, 44; surface, 43–44; ground, 44–46, 54

forest management agreements, xiii, 143
forests: montane coniferous, 169; northern hardwoods, 164, 167; old growth, 63–64
forestry, 142–43, 146–47, 149, 155, 158–59; clear-cuts, 142, 146–49, 163; deforestation, 142; ecologically sustainable, 141–42, 147, 151; herbicides, 158–59; paper mills, 46–47
forest-tundra, 5, 12, 167, 168, 170, 171
fossils, 28–29, 101, 105, 144
foxes: arctic, 15; red, 15, 17, 74, 76, 85, 86, 89, 96, 99
freeze-up of lakes or rivers, 117, 122
Fuller, William, 8, 154

Geist, Val, 51–52
geological regions, 25, 31
giens, 80
glacial landforms: drumlins, 35; erratics, 35; eskers, 36; kames, 36, 129; kettles, 36, 101, 121, 129; meltwater channels, 36; moraines, 35; roches moutonnées, 35; striations, 35; till, 35; troughs, 35
glacial processes: deposition, 35; erosion, 35; rebounding, 34
glaciers, 34; cold-based, 34; warm-based 35
global warming, 20–24, 34, 43, 111, 123. *See also* climate change
Great Slave Lake, 119, 122
greenhouse gases, 20, 111–12, 121
grouse, 18, 75, 76, 85, 94
Gulf Stream, 5, 156

hares: European, 87; mountain 87; snowshoe, 15, 75, 76, 77, 85–99. *See also* population cycle
Hudson Bay, 2, 23, 31; Lowlands, 32, 34
Hudson's Bay Company, xiii, 85, 90
hyaline cells, 106–7
hypothermia, 76

ice, 20, 22, 41–42, 71, 117; candling, 119
Ice Age, 21, 28, 33–34
indigenous people, 39–41, 49–50, 73, 79–83, 108, 109, 112–13, 120, 121, 144. *See also* First Nations
Innuitian Mountains, 32, 33
insects, 15, 22, 47–48, 74, 102–3, 108, 109, 114, 122, 123, 125, 133
Interior Plains, 32–33, 38, 112, 168, 169, 170
irruptions, 18

Keith, Lloyd B., 87, 91–92, 96, 98–99
Kluane National Park and Reserve, 7, 171
Koyukon, 144, 149
Krebs, Charles J., 92, 96–99
Kyoto Protocol, 23–24

Labrador, 2, 14, 120, 124, 143, 165
ladder ("torch" trees), 44, 53, 67
Lahti, Torbjörn, 153, 157–58
Lake Baikal, 121–22
lakes: boreal, 5, 22–23, 39, 41, 73, 110, 113, 115–19, 121–27, 135, 163; glacial, 36, 123; great, 122–23, 131, 133–34; shield-edge, 122–23; turnover of, 116–18
Lake Superior, 122, 134
Lapland, 146
Larsen, James A., 13–14
lead poisoning, 124–25
lichens, 13, 41, 44, 79
lichen woodland, 4, 12–14
loons: adaptations for diving, 131–32; Arctic, 130; breeding, 130, 133–36; common, 129–39; environmental threats, 133, 135; evolution, 129–30, 134; human interactions, 129, 132–33, 135–38; hunting of, 132–33; migration, 131, 133–34; nesting platforms, 137; Pacific, 130; predators, 136–37; red-throated, 130; storlom, 145–46; territories, 134–35; vocalizations, 129; yellow-billed, 130
lynx, Canadian, 15, 76, 85, 89–91, 92, 96, 98, 99. *See also* population cycle

Maine, 9, 110, 124
mammals, small, 15, 19, 74–75, 76
Manitoba, 17–18, 28, 34, 79, 110, 120, 143, 168
mercury, 133
Mesozoic Era, 28, 32
Michigan, 56, 110, 124
mining, 141, 143, 155
Minnesota, 9, 56, 105, 110, 137
Montana, 1, 9
moose, 48–51, 76, 111, 113, 130; habitat of, 49–51; moose-calling, 49
mosquitoes, 102, 108, 109, 114, 163
moss glacier, 104–5
museums, 9, 165, 167, 168
muskegs, 38, 44, 94, 105–9, 112–14. *See also* peatland
Mutch, Robert, 67
mycorrhizae, 11, 65

Nelson, Richard K., 144
Nero, Robert, 18
New Brunswick, 142, 166
Newfoundland, 2, 7, 21, 86, 124, 125, 131, 143, 154, 165–66
New Hampshire, 137, 138
New York, 9, 124
Northwest Territories, 10, 13, 28, 79, 119, 154, 170

Norway, 144, 156
Nova Scotia, 125
Nunavut, 10, 13, 103, 154, 170

Ontario, 21, 28, 32, 34, 110, 124, 125, 135, 142–43, 167–68
organic terrain, 45, 64, 104–5, 108
orogeny, 29, 32
Övertoneå, Sweden, 156–61, 164
owls, 18, 75, 85, 89, 92, 99

pack ice, Arctic, 22
Paleozoic Era, 28, 32, 33
palsa, 105
paludification, 64
passerines, 15, 17–19, 75
Peace-Athabasca delta, 120
peatland, 44, 104, 107, 110, 111, 116, 121, 146–47. *See also* muskegs
pelican, white, 19
permafrost, 5, 11, 22, 66, 104–5, 107
Pielou, E. C., 10–12
pines, 2, 41; eastern white, 51; jack, 9, 12, 43, 51–58, 60, 63; lodgepole, 43, 53, 55, 60
pinosylvan toxin, 94–95
plants, 15, 21, 23, 60–61, 64, 79, 83, 121, 161; avoiders, 61; endurers, 61; evaders, 60, 61; invaders, 60–61; resisters, 60–61
plants, carnivorous, 108, 109
plate tectonics, 29–30
Pleistocene, 28, 31, 33, 34, 37
poplar, balsam, 2, 9, 96
population cycle, 85–88, 96–99; stress and, 91–92, 97, 99; time lags in, 91, 95, 99. *See also* cycles
prairies, 1, 3, 22, 42
Precambrian Era, 28, 31, 33
Precambrian shields, 29, 31, 38, 112, 124. *See also* Canadian Shield
Prince Albert National Park, 18, 45, 64, 76, 77, 134, 153, 168
Prince Edward Island, 166–67
Pruitt, William O., Jr., 8, 10, 45, 73–75, 79, 80, 147
ptarmigans, 75, 76
push mills, 154

Quebec, 5, 14, 19, 32, 85, 110, 120, 124, 125, 126, 142, 143, 167
quin-zhee, 80–83, 84

raccoons, 136, 137
reindeer, 146, 157, 158
Riding Mountain National Park, 18
rivers, northern, 5, 22, 115–16, 119–21, 127, 156

rocks: igneous, 25, 27; sedimentary, 26, 29, 32; metamorphic, 27, 32
Rocky Mountains, 2, 3, 87, 169
Rowe, J. Stan, 53–54, 60–61
Ruisniemi, Maija, 161–62
Rupert's Land, xiii, 143
Russia, xiii, 1–2, 12, 22, 44, 80, 108, 110, 121–22, 147–49

Saint Lawrence Lowlands, 167
Sami, 146
Saskatchewan, 5, 18, 33, 70, 80, 83, 86, 110, 116, 120, 134, 143–44, 153, 163, 168–69
Scandinavia, xi, 5, 12, 80, 87, 104, 113, 124, 144–47, 151, 156–69
Schindler, David W., 23, 120, 121, 122, 141–42
Scotter, George W., 53–55
seed-bankers, 60, 61
Siberia, xiii, 1, 12, 22, 70, 87, 104, 119, 121–22, 124, 147–49; Cossacks of, 148
Sinclair, A. R. E., 92–93, 96
snow: insulation of, 72–73; metamorphism of, 72; sintering of, 72–73, 82; taiga, 70, 72–73, 79, 87; tundra, 72, 74, 80
snowflakes, 71–72; nucleating agent for, 70
snowshoes, 69, 79–80, 81
snow terminology: api, 73; kanik, 73; pukak, 74, 75; qali, 73–74, 75; upsik, 73
soils, 10–12, 36, 39–41, 44–46, 53, 60, 64, 65, 94, 102, 127, 121, 124, 125; spodosols (podzols), 11
solar radiation, 20–21, 34, 40–41, 93, 117
solstice, 70
Soviet Union, 148–49
Sphagnum, 44, 63–65, 101, 113–14, 146–47
spruce: black, 2, 9, 14, 43, 44, 51, 60, 63, 64–66, 94, 104, 105; white, 2, 9, 44, 61, 63, 67, 93, 96
squirrels: arctic ground, 89, 98, 99; red, 15, 54, 75, 89, 98
stromatolites, 28, 37

succession, plant, 43, 59–60, 63, 65–67, 94–95
Sun, angle of, 3, 9–10, 42
sunspots, 64, 93
Suomussalmi, Finland, 156, 157, 162–63, 164
Sweden, xiii, 113, 124, 138, 144–47, 149, 153, 156–61, 163–64

taiga, circumpolar, xiv, 112; defined, 1–2
tamarack, 9, 12, 41, 65
terpenes, 94–95
toboggan, 80
Tollund Man, 113
toxic metabolites, 94–95, 99
treeline, 2, 12–14, 104
trout farming, 110, 163
tundra, 1, 13, 14, 165, 167, 168, 170, 171

ventilation holes, 75
Vermont, 9, 87, 110, 124
Viereck, Leslie A., 53, 63

warblers, 18–19, 143
water balance, 42, 107
Weller, Gunter, 21–24
wetlands, 22, 102, 111, 121, 146–47
whooping cranes, 19–20, 150
Wihtiko, 112–14
wildflowers, 3, 60–61, 162
wild rice, 111, 163
Wisconsin, 87, 110
wolverines, 19
wolves, 15, 41–42, 47, 48, 76, 77, 150
Wood Buffalo National Park, 64, 77, 120

Yoho National Park, 28, 170
Yukon Territory, 5, 7, 13, 21–22, 29, 49, 92, 153, 171

zebra mussels, 23, 133
zones of subduction, 30, 32, 33
zones of vegetation, 3

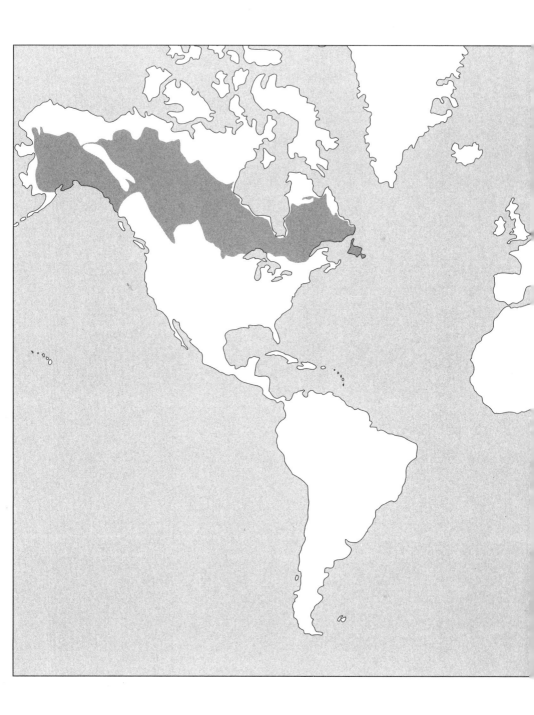